Synthesis Lectures on Engineering, Science, and Technology

The focus of this series is general topics, and applications about, and for, engineers and scientists on a wide array of applications, methods and advances. Most titles cover subjects such as professional development, education, and study skills, as well as basic introductory undergraduate material and other topics appropriate for a broader and less technical audience.

Jarred S. Glickstein · Soumyajit Mandal

Electromechanical Transmitters for ELF/VLF Radio

Springer

Jarred S. Glickstein
Case Western Reserve University
Cleveland, OH, USA

Soumyajit Mandal
Brookhaven National Laboratory
Upton, NY, USA

ISSN 2690-0300 ISSN 2690-0327 (electronic)
Synthesis Lectures on Engineering, Science, and Technology
ISBN 978-3-031-15122-4 ISBN 978-3-031-15123-1 (eBook)
https://doi.org/10.1007/978-3-031-15123-1

This Springer imprint is published by the registered company Springer Nature Switzerland AG
The registered company address is: Gewerbestrasse 11, 6330 Cham, Switzerland

To our families, without whose continuous support and encouragement this book would never have been written.

Foreword

Studying radio waves on the low-frequency side of the spectrum is an incredibly interesting endeavor. At High Frequencies (HF, 3-30 MHz), you have long-range communications and over-the-horizon radar thanks to reflection from the Earth's ionosphere, an electrically charged layer bordering the atmosphere and space environment. At Low Frequencies (LF, 300-3000 kHz), there are a number of communication beacons, and long-range communications are made possible by elevated diffraction over the curved Earth's surface (see Higginson-Rollins et al. 2017 for an example). At Very Low Frequencies (VLF, 3-30 kHz), you have natural impulsive signals from lightning, called sferics, that can be seen globally thanks to the "Earth-ionosphere waveguide", in addition to Navy beacons that are used, for example, to communicate with submerged submarines (see Barr et al. 2000 for a review). At Ultra Low Frequency (ULF, 0.3-3 kHz, as defined by the International Telecommunications Union), a number of interesting natural signals called whistlers, chorus, and hiss can be detected that serve as a window into the incredibly dynamic near-Earth space environment. At Super Low Frequency (SLF, 30-300 Hz, as defined by the International Telecommunications Union) and Extremely Low Frequency (ELF, 3-30 Hz, as defined by the International Telecommunications Union), there are harmonics from the power grid, and some energy from the "slow" components of lightning, as well as Schumann resonances which can be thought of as "around the world" resonances. Lower still you get oscillations in the Earth's magnetic fields, the ability to sense and detect underground structures, and so much more. Superimposed on top of all of this is a rich world of geoscience, in which everything from the Sun's x-ray emissions, to the space environment, to lightning and thunderstorms, all have something to say.

There are essentially three pillars to the excitement in all these low-frequency areas. (1) Generation, (2) Propagation, and (3) Detection and processing. The third pillar is fairly well developed. We have long had receivers in place that are sensitive enough to detect even weak VLF signals from many thousands of km away, for example (see Cohen et al. 2018 and Cohen et al. 2010, for example). Lots of low-frequency data are being collected and made available widely (see Cohen 2020 for example), and increasingly sophisticated techniques are emerging for wide data analysis using cloud storage, high performance computing, and machine learning. The second pillar is also well established, as the physics of low-frequency propagation in, below, and through the ionosphere and magnetosphere has long been mapped out (see Budden 1985, for example).

The first pillar, however, has long been a huge challenge. At HF, transmission is possible but often requiring an "antenna farm" to get decent directivity and efficiency. As you go lower in the spectrum, this gets harder and harder. At VLF, the existing facilities that are in the kW-MW range take 1000s of acres and have components the size of a building, yet still manage to broadcast only in a few hundred Hz bandwidth. In the SLF range, things get even worse, and as an example, consider that a 150 km long antenna operating in Wisconsin and North Dakota managed just 10 W or so of power. In these frequency bands, we are limited by the fact that wavelengths are very long. At 30 kHz, the wavelength is 10 km. So the traditional strategy of quarter-wave antennas is very difficult if not impossible. Electrically short antennas, on the other hand, are limited by the Chu-Harrington theory, which implies that you simply cannot have a small, efficient, and broadband antenna. At best you can have one of the three.

As exciting as low frequencies are, it is a big hindrance to not be able to easily generate our own signals. For example, we have to rely on lightning to generate broadband ELF-VLF emissions, and thunderstorms are entirely out of our control.

A number of efforts to circumvent this fundamental limit have been suggested (see Slevin et al. 2020 for an example), and in my view, all of them are worth a serious look. But if you think about it, the problem of radiating radio waves at long wavelengths is fairly simple intellectually. Just move the most amount of charge as far as you can, periodically, and you have an electric source. If you generate the highest amount of current encircling as big an area as you can, and you have a magnetic source. At long range, they look the same, so it doesn't really matter which one you use for some applications. In the near field, though the magnetic source will be much more effective at generating magnetic fields. Of course, in practice, it is very difficult to actually achieve this. This book will describe one such technique in which the mechanical motion of the antenna is used to generate either the electric or magnetic source. This is an alternative to antennas in which a conductor like a wire of metal is used to move charge around in the form of current. That's just another way to move charge and current around.

Certainly, the idea of mechanical motion to radiate will have challenges, but these challenges will be very different in nature than current and voltage antennas. In that sense, there may be areas where mechanical motion antennas can have an impact, particularly for high portable/compact needs where very long range is not needed.

Once a source is identified, the remaining exercise will be to construct a communication signal with a receiver and proper data analysis. In addition to communications in a man-made sense, these same treatments can be used to explore to a wide variety of geoscience topics, as well.

June 2022

Morris B. Cohen
Georgia Institute of Technology
Atlanta, USA

References

K. G. Budden (1985), "The Propagation of Radio Waves in the Ionosphere and Magnetophere", 1st edition, Cambridge University Press.

Morris Cohen, Ryan Said, Evans Paschal, Jackson McCormick, Nicholas Gross, Lee Thompson, Umran Inan, J. C. Chang (2018), Broadband Longwave Remote Sensing Instrumentation, *Review of Scientific Instruments*, 89, 094501.

Morris Cohen, Umran Inan, and Evans Paschal (2010), Sensitive Broadband ELF/VLF Radio Reception with the AWESOME Instrument, *IEEE Transactions on Geoscience and Remote Sensing*, 48, 1, 3-16.

Morris Cohen (2020), Returning lightning data to the cloud, *Eos*, 101.

Richard Barr, D. Llanwyn Jones, Craig Rodger (2000), ELF and VLF radio waves, *Journal of Atmospheric and Solar-Terrestrial Research*, Vol. 62, Number 17-18.

Preface

Wireless communications between the earth's surface and underground or deep-water facilities (mines, shelters, storage areas, tunnels, submarines, undersea cables, etc.) are of great practical importance in oil and gas exploration, search-and-rescue missions, biosensing, geophysics, and other applications. However, the range of such links is fundamentally limited by the rapid attenuation of electromagnetic waves within conductive media such as earth or seawater. The range can be extended by using the extremely low-frequency (ELF) or very low-frequency (VLF) bands, but extremely large transmit antennas are then required to obtain sufficient bandwidth, channel capacity, and radiation efficiency. In 2017, the U.S. Defense Advanced Research Projects Agency (DARPA) launched the "A MEchanically Based Antenna" (AMEBA) program, which aimed to address this fundamental challenge by developing miniaturized ELF/VLF transmitters based on the mechanical motion of permanent magnets or electrets (rather than the conventional approach, which utilizes current flowing within conductors).

This book describes our work on new miniaturized AMEBA-type ELF transmitters based on the mechanical motion of permanent magnets or electrets. By utilizing the very high energy density of modern ferromagnetic and ferroelectric materials, such "electromechanical transmitters" can provide much higher field generation efficiency than conventional antennas, thus enabling practical ELF wireless communications links. In particular, we focus on magnetic field (**B**-field) transmitters realized using rotating permanent magnets. Chapters 1–2 describe the fundamental challenges of ELF wireless links and provide a historical overview of the attempts that have been made to address these challenges. The text then focuses on the design and implementation of practical electromechanical ELF transmitters using rotating permanent magnets, which is an interdisciplinary subject that spans multiple research areas including electromagnetics (Chaps. 3–4), power electronics (Chap. 5 and 6), control systems (Chaps. 6 and 8), and mechanical design (Chap. 7). It also describes how such transmitters can be combined with low-noise ELF receivers and signal processing algorithms (Chap. 9) to realize complete ELF wireless communications links in challenging environments, such as the earth's crust (Chap. 10).

Finally, Chap. 11 summarizes our conclusions and discusses avenues for future work in this exciting research area.

Cleveland Heights, OH, USA Jarred S. Glickstein
Merrick, NY, USA Soumyajit Mandal
May 2022

Acknowledgements

The authors would like to thank Prof. Arjuna Madanayake of Florida International University and Prof. Seungdeog Choi of Mississippi State University for serving as collaborators on the AMEBA project. Special thanks to Prof. Morris B. Cohen of the Georgia Institute of Technology for taking time out of his busy schedule to write the foreword.

Many thanks go out to Michael, a master machinist who has been an enormous help in debugging rotating assemblies and a very patient advisor on all things mechanical. Special thanks also go out to Tim and Ben at Ohio Caverns (https://ohiocaverns.com) for not only allowing access to the caverns for field test experiments but also helping to move equipment into the cave.

Last but definitely not least, the authors would like to thank their long-suffering families and friends for providing invaluable emotional and gastronomic support during the writing of this book.

Contents

1 **Introduction** 1
1.1 Overview of the Book 1
1.2 Motivation: Radio Silence in Conductive Media 1
1.3 Properties of ELF Communications 3
References 5

2 **Historical Overview** 7
2.1 Development of ELF/VLF Communication Systems 7
2.1.1 ELF/VLF Communication with Electrical Antennas 7
2.1.2 Underwater Communication Modes 13
2.1.3 Cave Communication 16
2.2 Development of Electromechanical ELF/VLF Transmitters 18
References 19

3 **Electromagnetic Modeling** 23
3.1 Field Generation 23
3.1.1 Rotating Magnetic Dipole 23
3.1.2 Rotating Electric Dipole 24
3.1.3 Effect of Ground Planes 25
3.1.4 Comparison Between Magnetic and Electric Dipoles 25
3.2 Power Consumption 25
3.2.1 Shape of the Dipole 27
3.2.2 Distributed Transmitters 27
3.2.3 Dynamic Stability 28
3.2.4 Interactions Between Rotating Dipoles 29
References 30

4 **Design of ELF Communications Links** 31
4.1 Natural and Artificial Activity in the ELF Channel 31
4.1.1 Atmospheric Effects 32
4.1.2 Schumann Resonance 33
4.1.3 Animal Response 34

4.2 Characterization of the ELF Channel ... 35
4.3 System Design ... 38
4.3.1 Data Modulation ... 39
4.3.2 Minimizing Power Consumption ... 39
4.3.3 Simulation Results ... 42
4.3.4 Differential Coding ... 42
4.3.5 Bit Error Rate (BER) Analysis ... 44
4.3.6 Data Encoding ... 46
References ... 47

5 Open-Loop Prototype ... 51
5.1 Prototype Experimental Setup ... 51
5.2 Prototype Experimental Setup: The Completed Assembly ... 53
5.2.1 Transmitter Design ... 55
5.2.2 Receiver Design ... 55
5.3 Experimental Results ... 57
5.3.1 Field Strength Versus Distance ... 57
5.3.2 Wireless NLOS Data Transmission ... 58
5.3.3 Sample Data Encoding ... 60
5.3.4 Practical Message Transmission ... 62
5.3.5 Bit Error Rate (BER) Analysis of the Prototype ... 63
5.3.6 Power and Energy Consumption with CP-FSK and CF-FSK ... 66
5.3.7 Effects of Non-random Data ... 67
References ... 68

6 Design of a Closed-Loop Transmitter ... 69
6.1 Main Design Goals ... 69
6.1.1 Bandwidth Selection ... 70
6.2 A Model for the Mechanical System ... 70
6.2.1 The Fundamental Model ... 70
6.2.2 Constraining the System and Optimizing the Model ... 71
6.2.3 Constraining the Communication Domain ... 72
6.2.4 Constraining the Mechanical Domain ... 74
6.2.5 The Gasoline Engine Analogy to the Electric Motor ... 78
6.2.6 Summary of Constraints for Optimization ... 79
6.3 Torque Required to Drive Assembly ... 80
6.3.1 Torque Due to Bearing Power Loss ... 81
6.3.2 Modulation Torque ... 82
6.3.3 Static Torque Due to Drag ... 84
6.3.4 Summary of Torque Requirements ... 85
6.4 The Optimum Magnet ... 86

6.4.1 Sizing the Magnet from Available Torque 86
6.4.2 Real Size of the Magnet, A Boundary Value Problem 88
6.5 Modular Transmitter Design 92
6.5.1 Resolving the Field Generation Limit 92
6.5.2 BER Analysis of Optimum Magnet Design 96
References 101

7 Implementation of the Closed-Loop Transmitter 103
7.1 Mechanical Design 103
7.1.1 Selecting a Motor 103
7.1.2 Air Supply 108
7.1.3 Permanent Magnet 111
7.1.4 Shaft Coupling 112
7.1.5 Computer-Aided Design of the Mechanical Assembly 112
7.1.6 Critical Speed of the Rotating Assembly 115
7.2 A Three-Phase Motor Driver 117
7.2.1 Drive Type and Requirements 117
7.2.2 Motor Controller Board 120
7.2.3 Revised Motor Controller Board 122
7.3 Machining and Assembly 123
7.3.1 Alternative Methods for Centering the Magnet 124
7.3.2 Returning to the Bigger Picture 125
7.3.3 "Fit by Feel" 126
References 128

8 Transmitter Control System 129
8.1 System Overview 129
8.1.1 Hardware Test Bench 130
8.1.2 A Model for the Plant 132
8.1.3 Remarks 134
8.2 Digital Controller 135
8.2.1 Driving the Motor 135
8.2.2 Feedback and Speed Measurement 137
8.2.3 Data Collection 143
8.2.4 A Model for the Controller 144
8.3 Software Interface 148
8.3.1 Datatype Organization 150
8.3.2 Message Encoding 152
8.3.3 Sendmail: Real-Time Data Modulation 152
8.4 Experimental Performance Characteristics 153
8.4.1 Testing the Drive 153
8.4.2 Tuning the Controller 156

8.4.3 System Response and Stability Margins ... 158
8.4.4 Modulation Energy Explained ... 161
References ... 163

9 ELF Receiver Design ... 165
9.1 Overview of the Receiver ... 165
9.2 Background ... 165
9.2.1 On Detection and Measurement Sensitivity of Magnetic Fields ... 166
9.2.2 Proposed Detector Design ... 166
9.2.3 Related Work ... 167
9.3 Receiver Design ... 168
9.3.1 Magnetic Field Sensor ... 168
9.3.2 Low-Noise Front-End ... 171
9.3.3 Measured Characteristics ... 173
9.3.4 3-Axis Antenna Calibration ... 175
9.4 Portable Field Unit ... 177
9.5 Performance and Concluding Remarks ... 177
References ... 179

10 A Complete ELF Communication System ... 181
10.1 Field Setup: Transmitter ... 181
10.2 Results of Line of Sight Testing ... 183
10.2.1 Single Tone Transmission ... 183
10.2.2 Established CP-FSK Data Link ... 184
10.2.3 Maximum Link Distance ... 187
10.3 Cave Deployment ... 189

11 Conclusions and Future Work ... 195
11.1 Conclusions ... 195
11.1.1 Introduction to the AMEBA Concept ... 195
11.1.2 Implementation of an Open-Loop AMEBA Prototype ... 195
11.1.3 Design of a Closed-Loop AMEBA Transmitter ... 196
11.1.4 Implementation of the Closed-Loop Transmitter ... 196
11.1.5 Transmitter Control System ... 197
11.1.6 Design and Implementation of an ELF Receiver ... 197
11.1.7 A Complete ELF Communication System ... 197
11.2 Future Work ... 198
11.2.1 Mechanical ELF Transmitters ... 198
11.2.2 ELF Receivers ... 199
References ... 199

Appendix: Spectral Content of Modulated PWM Signals 201

Glossary 209

Index 211

Acronyms

ADC	Analog-to-Digital Converter
AMEBA	A MEchanically Based Antenna
AWG	American Wire Gauge
AWGN	Additive White Gaussian Noise
AXI	Advanced eXtensible Interface
BER	Bit Error Rate
BLDC	Brushless DC motor
BWT	Burrows-Wheeler Transform
CF-FSK	Continuous Frequency—Frequency-Shift Keying
CLI	Command-Line Interface
CP-FSK	Continuous Phase—Frequency-Shift Keying
CPM	Continuous Phase Modulation
DAQ	Data Acquisition
DARPA	Defense Advanced Research Projects Agency
DMA	Direct Memory Access
DNL	Differential Nonlinearity
DSP	Digital Signal Processor
DUT	Device Under Test
ELF	Extremely Low Frequency
EM	Electromagnetic
FCC	Federal Communications Commission
FET	Field-Effect Transistor
FM	Frequency Modulation
FOPDT	First-Order Plus Dead Time system model
FPGA	Field-Programmable Gate Array
FSK	Frequency-Shift Keying
GaN	Gallium Nitride
HPS	Hard Processor System
IC	Integrated Circuit
INL	Integral Nonlinearity

JFET	Junction Field-Effect Transistor
LNA	Low-Noise Amplifier
MFSK	Multiple Frequency-Shift Keying
MLSE	Maximum Likelihood Sequence Estimate
NLOS	Non-Line-of-Sight
OFDM	Orthogonal Frequency Division Multiplexing
PCB	Printed Circuit Board
PID	Proportional-Integral-Derivative controller
PIO	Parallel Input-Output
PLL	Phase-Locked Loop
PSD	Power Spectral Density
PWM	Pulse Width Modulation
RF	Radio Frequency
RLE	Run-Length Encoding
SNR	Signal-to-Noise Ratio
SoC	System-on-Chip
SPI	Serial Peripheral Interface
SQUID	Superconducting QUantum Interference Device
STFT	Short-Time Fourier Transform
UART	Universal Asynchronous Receiver/Transmitter

1 Introduction

1.1 Overview of the Book

Wireless communication systems, which use electromagnetic fields to manipulate, convert, transfer, and recover information, are fundamental in enabling emergent and life-saving technologies. However, such technologies frequently stall before seeing widespread adoption or use. Low transmit efficiency or signal reception creates barriers to use for even the most lucrative life-saving techniques. In this book, we present an example of such elusive applications and, using low-noise instrumentation and the synergy between electrical and mechanical devices, demonstrate methods whereby these barriers to use are eliminated.

Specifically, we focus on the transmitter portion of the communication system. We show that the replacement of a conventional electrical antenna with a mechanically rotated magnetic dipole allows wireless communication at extremely low frequencies (ELF) with orders of magnitude higher power efficiency than equivalent electrical antennas. To demonstrate a complete communication link, a state-of-the-art 3-axis portable ELF receiver is constructed. The receiver achieves a sensitivity that is observed to be below the noise floor of the Earth's atmosphere and is also used to evaluate and study the conditions of the ELF channel across various locations.

1.2 Motivation: Radio Silence in Conductive Media

Wireless communications between the earth's surface and underground or deep-water facilities (mines, shelters, storage areas, tunnels, submarines, undersea cables, etc.) are very difficult due to the short skin depth δ of electromagnetic (EM) waves within conductive media such as earth or seawater, which leads to high attenuation.

J. S. Glickstein and S. Mandal, *Electromechanical Transmitters for ELF/VLF Radio*,
Synthesis Lectures on Engineering, Science, and Technology,
https://doi.org/10.1007/978-3-031-15123-1_1

521545 1 En 1 Chapter-print ☑ TYPESET ☐ DISK ☐ LE ☑ CP Disp.:8/10/2022 Pages: xxx Layout: German_T5

Let us consider a plane EM wave propagating along the z-axis within a lossy medium characterized by a conductivity σ and complex permittivity $\epsilon = \epsilon' + j\epsilon''$, where the imaginary term ϵ'' arises due to the phase lag between the field and the polarization. A similar expression can be used for the complex permeability of the medium, i.e., $\mu = \mu' + j\mu''$. Most materials in nature are weakly magnetic, such that $\mu \approx \mu'$. A time-harmonic wave at a frequency $\omega = 2\pi f$ in such media can be characterized by a complex propagation constant γ, such that the amplitudes of both the electric and magnetic fields (**E** and **B**, respectively) vary $\propto e^{-\gamma z}$.

In general, we can write $\gamma = \alpha + j\beta$ where α and β are the real and imaginary components of γ, respectively. In a lossless medium, $\alpha = 0$ (i.e., γ is purely imaginary), such that the wave amplitude remains constant with z. In a lossy medium, $\alpha > 0$, such that the wave amplitude attenuates exponentially $\propto e^{-\alpha z}$ as it propagates.

A simple analysis (found in many EM textbooks) shows that α and β can be written as

$$\alpha = \omega\sqrt{\frac{\mu\epsilon'}{2}}\left[\sqrt{1+\left(\frac{\sigma_{eff}}{\omega\epsilon'}\right)^2}-1\right]^{1/2}, \quad \beta = \omega\sqrt{\frac{\mu\epsilon'}{2}}\left[\sqrt{1+\left(\frac{\sigma_{eff}}{\omega\epsilon'}\right)^2}+1\right]^{1/2}, \tag{1.1}$$

where the imaginary part of ϵ has been lumped into an effective conductivity term $\sigma_{eff} = \sigma + \omega\epsilon''$. In practice, most materials are either *low loss*, such that $\sigma_{eff}/\left(\omega\epsilon'\right) \ll 1$, or *good conductors*, such that $\sigma_{eff}/\left(\omega\epsilon'\right) \gg 1$. In these extreme cases, simplified versions of these expressions are applicable. Specifically, in the low loss case, we get

$$\alpha \approx \frac{1}{2}\sigma_{eff}\sqrt{\frac{\mu}{\epsilon'}}, \quad \beta \approx \omega\sqrt{\mu\epsilon'}. \tag{1.2}$$

In this case, the propagation constant β is unchanged by the losses (to first order), while the attenuation constant α is proportional to σ_{eff}. On the other hand, in good conductors, we get

$$\alpha = \beta = \sqrt{\frac{\omega\mu\sigma_{eff}}{2}}, \tag{1.3}$$

such that both α and β are proportional to $\sqrt{\sigma_{eff}}$. Also note that both quantities are also proportional to $\sqrt{\omega}$, which results in a strongly frequency-dependent propagation velocity (i.e., wave dispersion).

In either case, the rate of wave attenuation is often characterized by the *skin depth* $\delta = 1/\alpha$, which defines the depth at which the wave amplitude is attenuated by a factor of $1/e \approx 0.37$. For good conductors, δ is given by

$$\delta = \frac{1}{\alpha} = \sqrt{\frac{2}{\omega\mu\sigma_{eff}}}. \tag{1.4}$$

This relationship highlights the challenge of communication through highly conductive media such as rock or seawater, where the skin depth and therefore the ability of RF signals to pass through the material are significantly reduced. Fortunately, $\delta \propto 1/\sqrt{\omega}$ where ω is the frequency. Thus, extremely low-frequency (ELF) EM waves in the 3–3 kHz range can penetrate long distances in RF-denied environments [2]. In other words, ELF waves have relatively large skin depths in these materials, thus allowing ELF communication to have appealing applications in presently RF-denied environments.

For example, seawater has a typical conductivity of $\sigma \approx 5$ S/m, such that it behaves as a good conductor (i.e., satisfies the condition $\sigma_{eff}/\left(\omega\epsilon'\right) \gg 1$) for frequencies up to ≈ 1 GHz. Nevertheless, the resulting skin depth remains relatively large at ELF (e.g., $\delta = 7.1$ m at 1 kHz), which allows undersea communications to depths of ~30 m with reasonable transmit power levels [1]. Note that the use of such low frequencies is typically not regulated by government agencies, unlike for higher-frequency bands. For example, in the United States, the FCC does not regulate the EM spectrum below 9 kHz.

1.3 Properties of ELF Communications

The availability of portable low-power ELF wireless transceivers would immediately enable communications within such RF-denied environments by enabling bidirectional low-data-rate wireless links with the surface. Promising applications of such transceivers include oil and gas exploration, search-and-rescue missions (cave rescues including when there are significant bodies of water in the signal path underground), biosensing (e.g., of marine and underground lifeforms), underwater and underground Internet of Things (IoT) networks (e.g., for ocean pollution monitoring and soil sensing), inter-submarine and submarine-to-surface wireless links, underwater autonomous robotics, advanced marine biology experiments, geophysics (e.g., subsurface imaging for locating geological conductors, such as ore deposits) [3], and atmospheric science (e.g., studies of the ionosphere) [4, 5]. The resulting near-field links have $\vec{E}$ and $\vec{B}$ fields that propagate independently; in this book, we analyze and experimentally demonstrate so-called magneto-inductive (MI) digital wireless links [6] based only on $\vec{B}$ fields. This is because magnetic fields are much less shielded than $\vec{E}$ fields by dielectric objects in the medium [7].

However, near-field ELF links present several problems, including (i) limited bandwidth due to the use of low frequencies, which in turn limits channel capacity, and (ii) the need to use electrically small antennas due to device size constraints. Fundamentally, the Chu-Harrington limit [8–10] lower-bounds the quality factor (Q) of small antennas to $\sim (c/\omega a)^3$, where a is the radius of its enclosing sphere. As a result, the bandwidth and radiation efficiency of small dipoles both degrade rapidly with frequency (as $\omega^4 a^3$ and $\omega^{3/2} a$, respectively), while small loops are even worse [1]. Thus, extremely large transmit antennas are necessary for obtaining reasonable bandwidth, channel capacity, and radiation efficiency at ELF [11–13]. ELF receivers can avoid similar issues by relying on noise matching (rather than impedance matching) to get high sensitivity. In particular, broadband noise matching is possible even for high-Q source impedances [14], which allows ELF receivers with mod-

521545_1_En_1_Chapter-print ☑ TYPESET ☐ DISK ☐ LE ☑ CP Disp.:8/10/2022 Pages: xxx Layout: German_T5

estly sized antennas to obtain excellent sensitivity (< 0.1 fT/Hz$^{1/2}$) over a broad frequency range [15–17]. Thus, miniaturized and highly sensitive ELF receivers are available, while far-field transmitters remain physically large and extremely power-hungry. Fortunately, the issue of poor radiation efficiency can be avoided in the near field, where the radiated fields are small compared to the stored fields. In fact, various switching-based direct antenna modulation (DAM) schemes have been proposed for decoupling the efficiency, bandwidth, and Q of near-field antennas [18–20]. Hence, this book focuses on miniaturized near-field ELF transmitters, which are the key for realizing short-range (up to $\sim$1 km) bidirectional wireless links in conductive media.

Military ELF and VLF facilities that are designed to communicate at a few dozen bits per second data rate require antenna "farms" that are built on massive camp-like ground facilities and require hundreds of kW, even MW, of transmit power to generate the required levels of signal-to-noise ratio (SNR) at the receiver, which can be thousands of km away from the transmitters and located several hundred meters under the ocean surface. The idea of mechatronic antennas and modulation schemes can significantly reduce both the physical size and power consumption of such a transmit site.

Since near-field transmitters are not limited by antenna bandwidth or Q, the key issue for practical implementations is simply *field generation efficiency*, which is defined as the magnitude of $\vec{B}$ per unit power consumption at a given distance r, and has units of T/W. Modern ferromagnetic and ferroelectric materials (permanent magnets and electrets, respectively) have extremely high energy density and display strong permanent dipole moments, which makes them promising for achieving high field generation efficiency [2]. For example, let us compare a cylindrical NdFeB grade N52 magnet (remanent field $B_r = 1.5$ T, length = radius = 10 cm) with a N-turn current loop of the same cross-sectional area. In order for the two to generate the same static $\vec{B}$ field, the loop needs to carry a very large DC current of $\sim 10^5/N$ A. Hence, while the static magnet consumes no power, we estimate that even a tightly packed copper loop will consume > 4 kW. Superconducting coils would eliminate the power dissipation problem, but modulating their persistent current in order to transmit data would then become highly challenging.

These facts suggest a fundamentally new approach to ELF transmitter design in which oscillating $\vec{B}$ fields (similar to those from oscillating electrical currents inside conductors) are generated by the periodic mechanical motion of a magnet or electret. This all-mechanical transmitter approach is more power-efficient at low frequencies than conventional electrical transmitters, as analyzed in earlier work [1, 2]. It also differs from early electromechanical transmitters, which simply coupled high-frequency AC generators (e.g., the Alexanderson alternator [21]) to dipole or loop antennas. The remainder of this book will focus on such all-mechanical ELF transmitters and their use in realizing power-efficient wireless communications links in conductive media.

References

1. Madanayake, A., Choi, S., Tarek, M., Dharmasena, S., Mandal, S., Glickstein, J., Sehirlioglu, A. (2017). Energy-efficient ULF/VLF transmitters based on mechanically-rotating dipoles. In *2017 Moratuwa Engineering Research Conference (MERCon)* (pp. 230–235). IEEE.
2. Tarek, M. T. B., Dharmasena, S., Madanayake, A., Choi, S., Glickstein, J., Liang, J., Mandal, S. (2018). Power-efficient data modulation for all-mechanical ULF/VLF transmitters. In *2018 IEEE 61st International Midwest Symposium on Circuits and Systems (MWSCAS)* (pp. 759–762). IEEE.
3. Singh, A., & Sharma, S. P. (2016). Interpretation of very low frequency electromagnetic measurements in terms of normalized current density over variable topography. *Journal of Applied Geophysics, 133*, 82–91.
4. Inan, U. S. (1990). VLF heating of the lower ionosphere. *Geophysical Research Letters, 17*(6), 729–732.
5. Hayakawa, M. (2007). VLF/LF radio sounding of ionospheric perturbations associated with earthquakes. *Sensors, 7*(7), 1141–1158.
6. Domingo, M. C. (2012). Magnetic induction for underwater wireless communication networks. *IEEE Transactions on Antennas and Propagation, 60*(6), 2929–2939.
7. Hansson, K. Greenebaum, B. (2018). Environmental and occupationally encountered electromagnetic fields. *Bioengineering and Biophysical Aspects of Electromagnetic Fields*, 1–34.
8. Chu, L. J. (1948). Physical limitations of omni-directional antennas. *Journal of Applied Physics, 19*(12), 1163–1175.
9. Harrington, R. F. (1960). Effect of antenna size on gain, bandwidth, and efficiency. *Journal of Research of the National Bureau of Standards-D. Radio Propagation, 64*(1).
10. Wheeler, H. (1975). Small antennas. *IEEE Transactions on Antennas and Propagation, 23*(4), 462–469.
11. Rowe, H. (1974). Extremely low frequency (ELF) communication to submarines. *IEEE Transactions on Communications, 22*(4), 371–385.
12. Field, E. C., Kies, L. R., Bannister, P. R., Ingram, R. F., Hopkins, W. D., & Roberts, M. A. (1989). An aerostat-supported ELF/VLF transmitter. *Radio Science, 24*(02), 235–246.
13. Barr, R., Ireland, W., Smith, M. J. (1993). ELF, VLF and LF radiation from a very large loop antenna with a mountain core. In *IEE Proceedings H (Microwaves, Antennas and Propagation)* (Vol. 140, No. 2, pp. 129–134). IET Digital Library.
14. Mandal, S., Utsuzawa, S., Cory, D. G., Hürlimann, M., Poitzsch, M., & Song, Y. Q. (2014). An ultra-broadband low-frequency magnetic resonance system. *Journal of Magnetic Resonance, 242*, 113–125.
15. Smith, A. J. (1995). VELOX: A new VLF/ELF receiver in Antarctica for the global geospace science mission. *Journal of Atmospheric and Terrestrial Physics, 57*(5), 507–524.
16. Harriman, S. K., Paschal, E. W., & Inan, U. S. (2009). Magnetic sensor design for femtotesla low-frequency signals. *IEEE Transactions on Geoscience and Remote Sensing, 48*(1), 396–402.
17. Cohen, M. B., Inan, U. S., & Paschal, E. W. (2009). Sensitive broadband ELF/VLF radio reception with the AWESOME instrument. *IEEE Transactions on Geoscience and Remote Sensing, 48*(1), 3–17.
18. Xu, X., Wang, Y. E. (2007). Beyond the efficiency bandwidth limit with switched electrically small antennas. In *2007 IEEE Antennas and Propagation Society International Symposium* (pp. 2261–2264). IEEE.
19. Azad, U., & Wang, Y. E. (2013). Direct antenna modulation (DAM) for enhanced capacity performance of near-field communication (NFC) link. *IEEE Transactions on Circuits and Systems I: Regular Papers, 61*(3), 902–910.

20. Özdemir, T., Davis, C. N., Ariando, D., Mandal, S. (2020). Direct antenna modulation for wide-band HF communications. In *2020 IEEE USNC-CNC-URSI North American Radio Science Meeting (Joint with AP-S Symposium)* (pp. 7–8). IEEE.
21. Phillips, V. J. (1997). The Alexanderson alternator. *Engineering Science and Education Journal, 6*(1), 37–42.

2 Historical Overview

2.1 Development of ELF/VLF Communication Systems

2.1.1 ELF/VLF Communication with Electrical Antennas

Immense ground facilities and megawatts (MW) of power are needed to build long-distance transmitters for typical ELF applications, such as surface-to-submarine low-bandwidth teletype links [1–3]. For example, the powerful Navy VLF transmitter NAA (24.0 kHz) occupies several km^2 of land in Cutler, Maine, and broadcasts up to 1.8 MW with a radiation resistance of $R_r \approx 0.17\ \Omega$, radiation efficiency $\eta_{rad} \approx 65\%$, and bandwidth $BW \approx 100$ Hz. Moreover, it is impossible to transmit in the opposite direction (submarine-to-surface) because of the large antenna sizes and transmit power levels. As a result, ELF communication between the surface and underwater or underground facilities is essentially one-way. Nevertheless, the need for wireless communication with submarines has motivated at least four countries (the United States, Russia, India, and China) to construct such immense facilities.

2.1.1.1 Project Sanguine

In the 1960s, the U.S. Navy initiated Project Sanguine, a proposed ELF transmitter hardened against nuclear attack and intended to enable one-way wireless communication with submarines at depths where they remain undetectable. The intended implementation involved enormous antennas buried across ~50% of Wisconsin, though concerns over environmental impact and lack of funding limited the implementation. Project Sanguine later devolved into two smaller base stations implemented under a new name, Project ELF [4].

Figure 2.1 shows the location of both Project ELF base stations, which utilize a combined total 84 miles (135 km) of transmission lines to form dipole antennas. Each base station uses two antennas oriented orthogonally for omnidirectional transmission. The transmitter

J. S. Glickstein and S. Mandal, *Electromechanical Transmitters for ELF/VLF Radio*,
Synthesis Lectures on Engineering, Science, and Technology,
https://doi.org/10.1007/978-3-031-15123-1_2

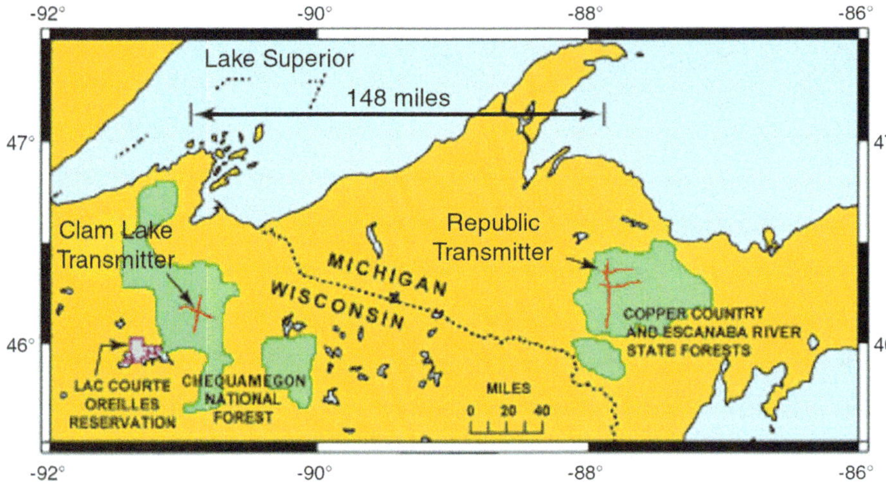

Fig. 2.1 Location of Clam Lake and Republic transmitters in Wisconsin, USA, which comprise Project ELF

operates at a center frequency of 76 Hz and consumes a nominal 2.6 MW. At full operational capacity, the radiated power in the ELF field is only 8 W.

The U.S. military still operates these two ELF transmitters, one of which is located near the town of Clam Lake and was constructed in the late 60s (ca. 1969) [5]. The initial project specifications were to bury enormous antennas. Due to lack of funding, Project ELF was finally implemented using above-ground dipole antennas. The site of the transmitter at Clam Lake is shown in Fig. 2.2. Compared with the base station and power generation, the transmitting antenna (a ground dipole) extends far into the surrounding forest, spanning 14 miles (23 km) and buried up to 6 ft underground [6].

Compared with power generation and transmitter modulation, the land area requirement for the dipole antennas is large. Figure 2.3 depicts the layout of one ground loop antenna at the Clam Lake base station. The second (orthogonal) antenna is not shown. Project ELF was decommissioned in 2004.

2.1.1.2 VLF Transmitter Cutler

As mentioned earlier, the Cutler Naval Station NAA in Cutler, Maine, includes a VLF transmitter that operates at 24 kHz. The transmitter requires 2 MW and operates at nominally 50% efficiency, radiating 1 MW. As a VLF antenna, transmit range is limited compared with Project ELF; however, the radiation efficiency is far greater. The Cutler station communicates one way with the U.S. Navy's Atlantic fleet.

The site map for the Cutler Naval Station is shown in Fig. 2.4. Complete implementation of the transmitter at the station uses two electrical antenna arrays. Most of the station layout

Fig. 2.2 Aerial view at the site of Project ELF transmitter in Clam Lake, Wisconsin

Fig. 2.3 Cartoon depicting the transmitting antenna and power/control station used at Clam Lake

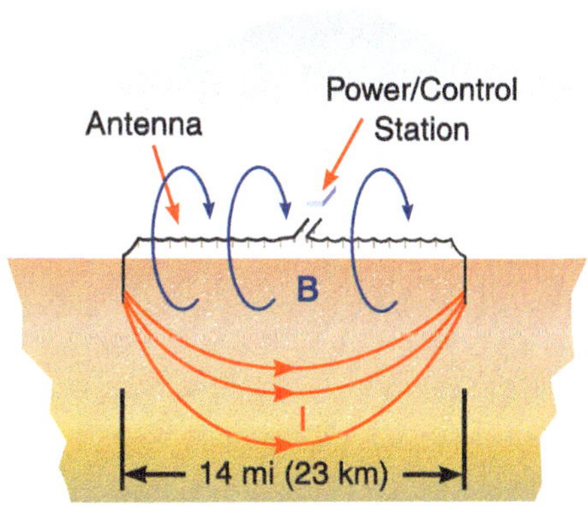

is devoted to supporting these antennas. A total of 26 towers distributed around the central transmitter building are used to support the massive antennas (13 per antenna). The towers are nearly 1000 ft high (see Fig. 2.5).

Since the transmitter was commissioned in 1960, maintenance has been a constant. The station continues to operate to date.

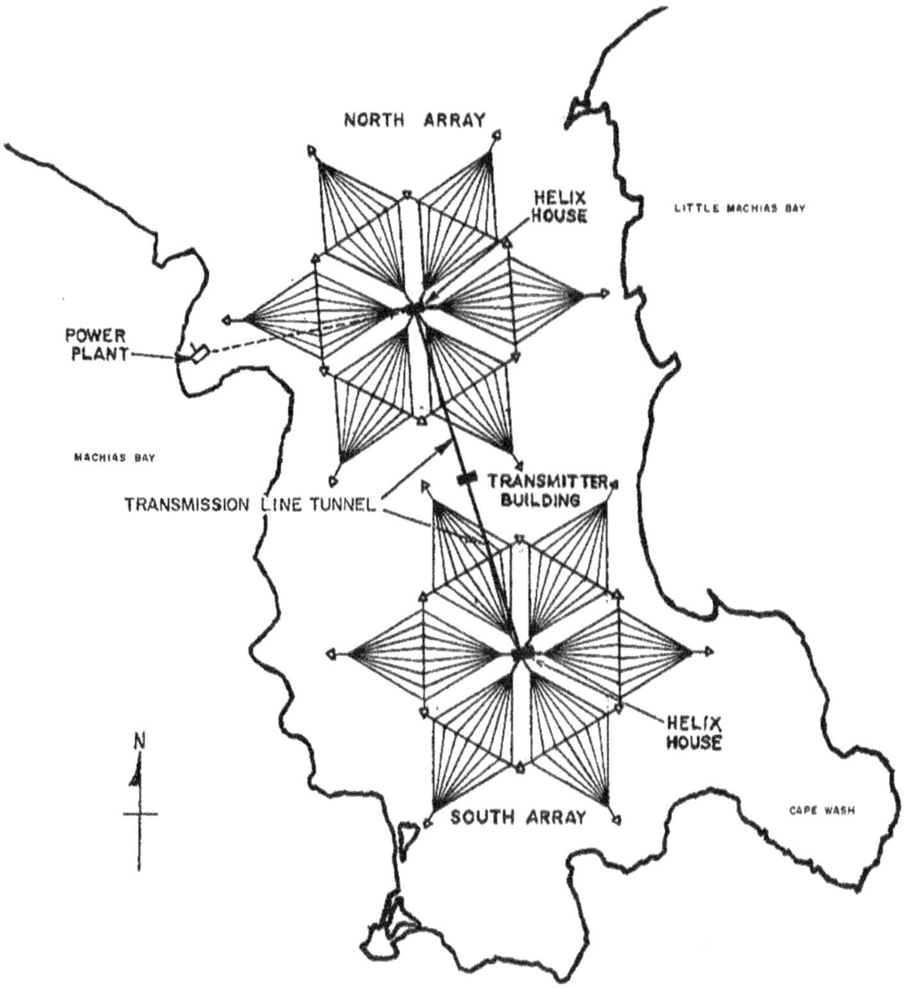

Fig. 2.4 Site plan drawing of the VLF transmitter at Cutler Naval Station in Cutler, Maine

2.1.1.3 Naval Communication Station Harold E. Holt

The Naval Communication Station *Harold E. Holt* was commissioned (historically under various names) in 1967, located on the North West Cape peninsula of Australia. The U.S. government leased land in this region of the globe to construct a VLF transmitter to communicate with navy vessels beyond the range of existing stations on U.S. soil. The station's VLF transmitter operates at 19.8 kHz, consuming 1 MW of power. If operating at 50% efficiency as reported for VLF Transmitter Cutler, the radiated power is 0.5 MW.

Thirteen radio towers stretching an area of ∼4 square miles form a Trideco array of the type used at VLF Transmitter Cutler. A drawing of this layout is shown in Fig. 2.6.

Fig. 2.5 Aerial view of Cutler Naval Station looking toward the southern shoreline

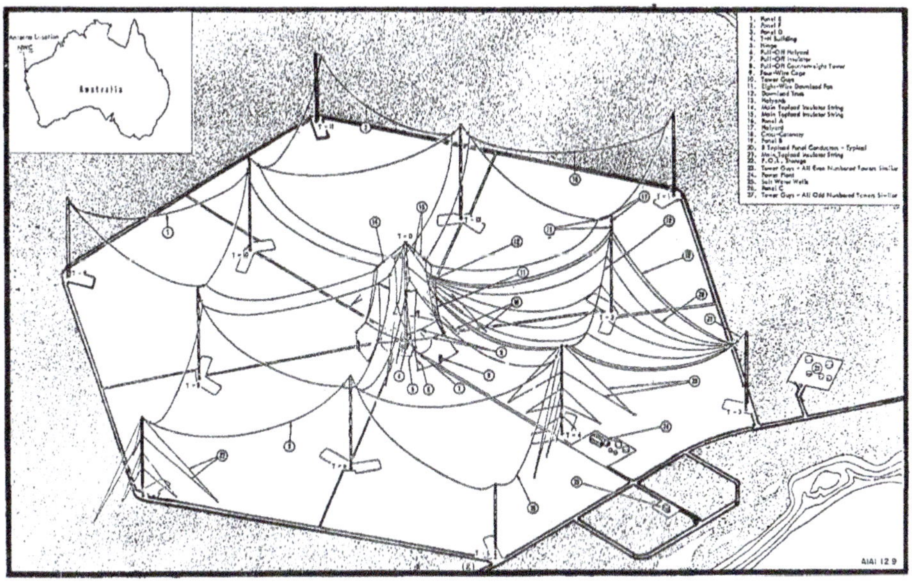

Fig. 2.6 Drawing of antenna layout at Naval Communication Station Harold E. Holt

Buried copper mesh under the entire array forms a ground plane. The use of this transmitter continues to the present day. Since 1974, the United States Navy and Royal Australian Navy have established joint operation of the transmitter.

2.1.1.4 Other Known ELF/VLF Transmitters

Four countries are known to have constructed ELF transmitters. Though details are not always freely provided by their operators, ELF transmitters using electrical antennas are by their nature remarkably difficult to conceal.

For example, while there is no direct public information available regarding the ZEVS station transmitter used by the Russian Navy, a massive antenna array is visible on satellite images in the Kola Peninsula of north-western Russia. It has been closely studied by amateur radio enthusiasts of sub-22 kHz communication, who quickly become so fascinated by such reports that they have pinpointed this antenna as the source of a 82 Hz radiating magnetic field. This field has been observed in Norway [7] and Italy [8]. Transmit power is estimated to be within 10–14 MW.

A spectrogram of the ELF channel showing a message transmitted by ZEVS is produced in Fig. 2.7. Power-line interference is also observed at 50 and 60 Hz, and Schumann resonances are visible. Romero reports that the latter is due to long exposure (40 min) [8].

The Indian Navy constructed the INS Kattabommam VLF transmitter in 1990. The transmitter is located in Vijayanarayanam, Tamil Nadu, and spans 4.7 square miles. The frequency of operation and power consumption are unknown.

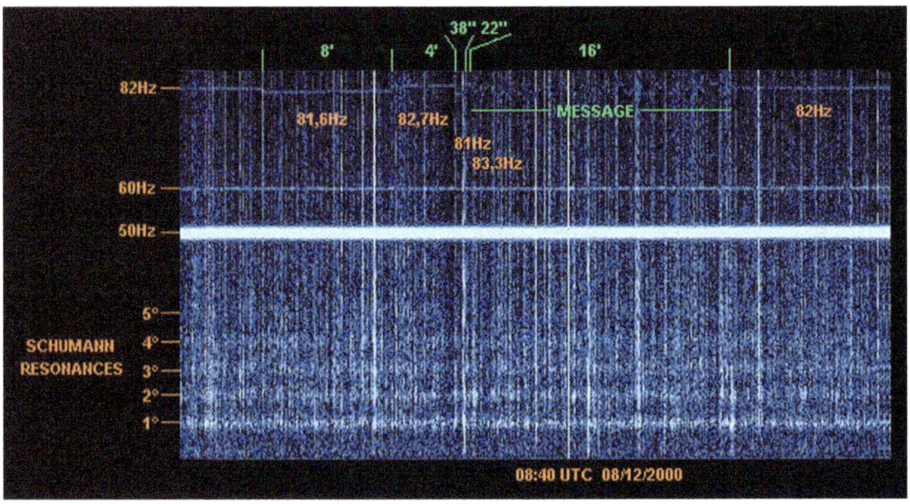

Fig. 2.7 Message transmission and carrier detected from ZEVS transmitter 82 Hz. Image reproduced with permission from [8]

China has recently commissioned an ELF transmitter occupying 1400 square miles[1] known as WEM. It is known to operate at 300 Hz though specific details on the transmitter are limited [1]. It is referenced in the literature as a tool for earthquake detection and mineral prospecting in addition to military communication [1, 2]. Recent media publications have speculated on the purpose and safety of WEM [3, 9]. This reaction, without details to form a basis for such speculation, may partially explain why there is limited public information available.

2.1.2 Underwater Communication Modes

Underwater communications are applicable to environmental exploration, personal recreational activities, military defense operations, and the oil industry. Beyond RF electromagnetic communication, acoustic and optical modes of wireless communication are standard in underwater operations. We also see extensive use of underwater communication systems for mapping, ranging, and sensing. All are exceptional examples of how communication systems are used to study channel dynamics through their effects on message transmission, as originally analyzed by Claude Shannon [10].

Acoustic waves: Communication systems for direct source-to-destination transfer of information in the underwater acoustic channel gain less attention than those used for underwater localization and sound signal analysis [11].

Sound navigation and ranging (SONAR) is used for imaging and source localization in aquatic environments. SONAR is typically operated above 50 kHz for military and commercial purposes, while low-frequency (1–50 kHz) systems are also used by both commercial and private individual operators at short range. Acoustic communication through seawater uses portions of the low-frequency region shared with SONAR and can have a nominal range of 10–100 km for transmitting frequencies 8–10 kHz [12]. At frequencies below 500 Hz, acoustic waves have been detected at distances of megameters away from their source [13]. In the 1950s, the United States Navy implemented a sound surveillance system (SOSUS) using arrays of microphones on the sea floor to track submarines. The SOSUS system was later repurposed to track and listen to whales up to 1700 km km away from the nearest microphone [13].

Passive acoustic surveillance systems are desirable for their improved stealth (for military and security purposes) and ability to collect information without disturbing the natural habitat of sea creatures (for biological studies). These systems have seen significant advancement since the days of their early development. In recent work, Agarwal et al. show that multiple wideband source localization is possible using a single 3-axis acoustic sensor [14].

Propagation of acoustic waves is slow; the speed of sound in seawater is only (nominally) 1500 m/s. The channel is nonuniform, with biological noise from fish and geological noise

[1] This is approximately $\sim 5\times$ the size of New York City.

from moving currents. On one side, the channel is bounded by the interface with the sea floor, and on another, it is bounded by moving surface waves. These effects create Doppler spread dependent on the velocity of the source, receiver, and moving components in the channel. Variance in channel conditions also introduces frequency-dependent propagation loss, and the slow propagation and varied speed of sound across the nonuniform channel create delays in multipath propagation modes [12, 15].

To summarize, acoustic waves have relatively low absorption in the underwater environment and are effective for bidirectional short- and long-range links. Underwater communication systems using acoustic transducers are thus quite common and many options are readily available as off-the-shelf commercial products [15]. These communication systems are used by military submarines for *ad hoc* communication with other vessels and nearby divers. Manned and autonomous underwater vehicles use acoustic data links during expeditions for exploration, research, and in search-and-rescue contexts. Applications also include remote operation and control of sensing equipment for deep sea fishing, off-shore oil exploration, and environmental monitoring. Historically, these systems were limited to low (1–100 b/s) data rates. Present day small form factor devices for fully submersed *ad hoc* communication provide data rates in excess of 10 kb/s over long (>1 km) distances [15]. Several hundred kbps can be achieved over short (1–3 m) distances [16].

Electromagnetic (EM) waves: EM waves have historically had limited usefulness in underwater communication systems due to their high attenuation in conductive media such as seawater; at present, there are two applications in which they are used. The first is unidirectional communication from one of the large land-based transmitters described in Sect. 2.1.1 to military submarines. The second is for short-range *ad hoc* wireless communication between divers and for sensor data collection, both limited to a range of nominally 1 m in known implementations [17]. This implementation is also called a non-contact link and is useful in place of wired cables when short-range data links are needed and it is possible to take advantage of the benefits of EM waves over acoustic or optical media. Such systems are a recent development and see relatively little use due to functional limitations and the already widespread adoption of acoustic transducers as the golden standard in underwater communication systems. **Given the present state of the art, the hybrid electromechanical ELF transmitters (i.e., AMEBAs) described in this book may disrupt this standard.**

Limitations in the prior art have prevented the widespread adoption of EM communication for long-range underwater data links, though it is poised to offer superior disturbance rejection and interference immunity over both optical and acoustic systems. Electromagnetic radiation at suitably low frequencies can penetrate solid barriers. It is generally robust against turbidity and disturbance effects of suspended plant matter and immune to ambient acoustic noise. Despite potential benefits, historical interest in EM data links and propagation through seawater is limited. Exploratory work by Siegel and King demonstrated (circa 1973) theoretical predictions and experimental confirmation of electromagnetic wave propagation between antennas submerged in seawater and spaced up to 30 m apart [18]. Test frequencies of 14 MHz and 100 kHz were used. Later work (circa 2004) by Al-Shamma'a et al. contributed similar findings [19].

Optical waves: Optical waves are useful for communication in shallow waters and over short (10–100 m) distances. While optical waves do not suffer from much attenuation [20], the data link requires a line of sight (LOS). Thus, the communication channel is strongly affected by scattering and absorption caused by turbidity. The channel is completely disabled if the LOS is lost by interference (e.g., crippled by a passing dolphin or fish) and sensitive to temporal variation in water parameters potentially caused by environmental phenomena, pollution, etc. Solar illumination of the water can also pose a problem by potentially creating excessive channel noise [21].

Though implementation of underwater optical communication systems has been historically limited in scope, a resurgence of interest has recently occured in response to new developments in distributed sensor networks including shallow water monitoring systems [22]. Compared with acoustic communication, the energy efficiency of optical data links is higher [16]. Optical waves also do not suffer from multipath propagation problems in shallow water and the speed of light permits much higher data rates and lower latency. In recent work, Wang et al. have demonstrated a multicarrier data link with a net data rate of 312 Mbps over a 21 m channel [23]. Spagnolo et al. suggest the state of the art has a theoretical data rate over 1000 Mbps and link distance over 500 m [24]. Note that while the authors of [24] do not specify this, these metrics are likely mutually exclusive. The state-of-the-art noise immunity techniques which enable such high data rates started appearing in the literature circa 2019. For example, in [25], Li et al. present techniques to compensate for temporal dispersion effects caused by water turbulence.

2.1.2.1 Recent Advances

Despite widespread adoption, underwater acoustic communication is generally recognized as one of the most difficult communication media in use today [26]. While existing systems have fulfilled the needs of ad hoc underwater communication, wireless transmission through the air-water interface at the surface remains an area in need of advancement. ELF radio waves propagate easily through this barrier, though existing transmitters capable of this feat are quite large as previously elucidated in Sect. 2.1.1. In underwater acoustics, the air-water barrier is usually treated as a perfectly reflecting pressure-release surface. Anecdotal evidence to the contrary has historically been dismissed and not gained much attention [27].

Recently some novel techniques have emerged that can cross this boundary through hybrid approaches, though ranging is limited and the technique has not been shown over a range suitable for industrial use. Wang et al. have proposed a thermoacoustic method for sk-to-underwater communication [28] and show how microwave-induced heating of seawater causes the emission of acoustic waves due to thermal expansion. Demonstration using a 3 GHz 4 kW microwave transmission was limited to 10 cm above and 6.5 cm below the water. Theoretical analysis by Wang et al. suggests that a 6.6 kW microwave transmission at

6 GHz could realize one-way communication up to 100 m away from the air-water boundary in either direction. Describing the motivation for this work, Wang remarks that the "massive and cumbersome equipment" required for ELF communication makes it unsuitable and other methods are needed. A later paper by Tonolini and Adib [29] presents a translational acoustic-RF communication system that enables one-way communication from underwater to air. The authors show that a radar can detect sound vibrations at the water surface produced by acoustic waves, with a data rate of up to 400 bps. Bidirectional communication by Long et al. [30] combines the previously developed techniques. The authors of [30] report a successful link between a surface device 1 cm above the air-water barrier and a submerged device 10 cm below.

2.1.3 Cave Communication

Effective communication is mandatory for safe exploration, surveying, and mining operations underground. Communication systems for use in these areas must be effective during standard operation and robust in emergency scenarios. There is no single standard underground communication system as the many underground environments formed by complicated patterns of geological conductors require the approach to be targeted for use in the specific environment. This has been a focus of research spanning over 50 years [31, 32].

Speleology

Speleologists need to communicate with the surface during expeditions into caves for surveying purposes, which often requires the design and implementation of an ELF communication system (not only the transmitter) suitable for use in the particular environment in advance of the expedition. Cocklebiddy Cave in Cocklebiddy, Western Australia, is 90% underwater. In an exploration of Cocklebiddy Cave over 55 h, Allum assembled two transceivers each with 32 m diameter coils of two turns and established an EM link between the surface and an unsubmerged portion of the cave 70 m below the surface [33]. Reid demonstrated the effectiveness of coupled radio transceivers to form a wired link between wireless radio domains above and below the surface in emergency rescue scenarios [34]. Similar scenarios are frequently repeated based on adapting needs.

Cave formations are alluringly unique, a feature that complicates the underground wireless communication channel as effectively as it attracts spelunkers to appreciate their natural beauty. The composition of rocks and minerals surrounding the cave and the distribution of passageways varies greatly by region, age, and (in some cases) modifications by human intervention.

Geologists and historians involved in cave exploration find a short supply of available communication systems. Below is an advertisement taken out by a professor of karst geology in an issue of *Speleonics* ca. 1982:

Needed, A working Cave Radio.

I need to purchase a cave radio system as soon as possible.

The needs of such exploratory researchers have been met by amateur radio enthusiasts with a deep interest in cave exploration, self-identified "cavers" who adapt radio communication systems to specific environmental circumstances.

Entering unexplored caves is undesirable due to the chance that human presence could damage historically significant artefacts or may be entirely unsafe. However, sometimes manned exploration becomes necessary. The need for effective geolocation and communication systems during surveying expeditions of such unexplored environments is limited by the large variance in skin depth of various minerals and mineralized waters [35]. Implementing sensor networks to monitor geological activities also requires customized solutions to optimized data links, which vary by location [36]. Such data links are also required while directly entering previously-explored caves, which is the general approach for amateur and non-professional cave explorers.

Mining

Miners rely on a variety of communication methods to maintain safe communications with the surface. However, the robustness of established communication systems is limited by the harsh environment of underground structures. No one solution addresses all needed communication requirements [37–39]. Distributed wired telephone networks with redundant interconnection are standard fixtures for voice communication, as are trolley carrier phones which communicate with surface-based personnel through electrically conductive trolley rails. The latter is also applicable for use as a hoist rope carrier, and in an emergency can be coupled to any electrically conductive single wire link between the miners underground and the surface. Wireless communication between miners can be implemented in parallel using a leaky feeder, a local wireless communications system based on massive networks of coaxial cabling which are intentionally unshielded to some RF frequencies and thus allow RF to "leak" and be detected by miner radios. These systems require distributed transceiver-repeaters to provide power to the entire antenna formed by the cabling. Fixed backbone technologies such as the former examples are fragile in emergency situations yet remain standard as primary communication modes due to general reliability in normal operating conditions.

In the mining industry, the robustness of existing communication networks in the event of an emergency is considered exceedingly poor [40]. Fires, flooding, explosions, and cave-ins present a hostile and changing environment in the event of an emergency. In recent work, wireless communication to the surface through distributed wireless nodes has been explored [41]. The implementation proposes the use of many distributed nodes similar to a leaky feeder system, except it can maintain a communication link when decoupled from direct wired connection using batteries as backup power sources and establishing a mesh link through multiple routes in the event a node is destroyed. Wireless communication in exposed openings from the surface to miners, as well as through the caves, has been a subject of recent interest to enable affordable, readily available communication through cave networks.

Effective emergency communication also depends on maintaining a reliable connection to the surface. This portion of the communication system is frequently maintained through a wired modality and remains a fragile part of the data link. Composition and structure of the cave environment form a waveguide which may be useful to carry particular frequencies through cave passages [32]. Radio communication in UHF bands used by conventional surface radio is potentially applicable if the propagation environment is well understood [42]. However, theoretical and experimental findings seem to have found no practically applicable approach to model such systems in the general sense.

Search-and-rescue missions

Underground search and rescue requires custom deployment in arbitrary environments, for which the present state of the field has no conclusive solution.

2.2 Development of Electromechanical ELF/VLF Transmitters

Electromechanical ELF transmitters have received significant academic attention since the concept was first described in 2017 as part of the DARPA AMEBA program [43]. Prototypes of transmitters based on rotating magnetic dipoles were demonstrated in [44, 45]. In both cases, the authors measured near-field antenna patterns and dependence of signal strength versus distance, but did not discuss data transmission. Other work has demonstrated all-mechanical very low-frequency (VLF) transmitters by using piezoelectric elements to produce oscillating electric fields [46, 47]. The generated VLF electric fields have either directly been used to transmit information [46] or coupled to a magnetostrictive material to generate a magnetic field [47]. These works again pay a great deal of attention to the design of the proposed antenna but do not consider data transmission. In the design of useful mechanically based antennas, an emphasis should be placed on suitable data modulation and encoding schemes and matters of practicality such as size and portability. These factors influence the design requirements for ELF transmitters based on mechanically rotating dipoles, which will be discussed in later chapters.

References

1. Di, Q., Xue, G., Yin, C., & Li, X. (2020). New methods of controlled-source electromagnetic detection in China. *Science China Earth Sciences, 63*(9), 1268–1277.
2. Guo, Z., Xue, G., Liu, J., & Wu, X. (2020). Electromagnetic methods for mineral exploration in China: A review. *Ore Geology Reviews, 118*, 103357.
3. Chen, S. (2018). China's new antenna is five times the size of New York City, but some fear it could be a cancer risk. *South China Morning Post*. Retrieved January 17, 2021, from https://www.scmp.com/news/china/science/article/2180071/chinas-new-antenna-five-times-size-new-york-city-it-also-cancer.
4. Grissett, J. D. (1980). Biological effects of electric and magnetic fields associated with ELF communications systems. *Proceedings of the IEEE, 68*(1), 98–104.
5. Genge, M., & Carlson, R. (1984). Project ELF electromagnetic compatibility assurance program-A status report. *IEEE Journal of Oceanic Engineering, 9*(3), 143–153.
6. Merrill, J. (1974). Some early historical aspects of project sanguine. *IEEE Transactions on Communications, 22*(4), 359–363.
7. Jacobsen, T. (2021). ZEVS, The Russian 82 Hz ELF Transmitter. ALFLAB. Retrieved July 27, 2021, from http://www.vlf.it/zevs/zevs.htm.
8. Romero, R. (2021). Radio Waves below 22 kHz. Open Lab. Retrieved July 27, 2021 from, http://www.vlf.it/.
9. Bhat, V. (2020). Unearthing China's communication efforts to assist long-distance nuclear submarines. *India Today*. Retrieved January 17, 2021, from https://www.indiatoday.in/news-analysis/story/china-communication-efforts-nuclear-submarines-1737292-2020-11-02.
10. Shannon, C. E. (1948). A mathematical theory of communication. *The Bell System Technical Journal, 27*(3), 379–423.
11. Anowar, M. T., Khan, M. N. H., Alam, M. M., Kabir, M. A., Hossen, M. D., Zahan, M. S., et al. (2016). A survey of acoustic underwater communications and ways of mitigating security challenges. *International Journal of Research in Engineering and Science (IJRES), 4*(6), 43–51.
12. Istepanian, R. S., & Stojanovic, M. (Eds.). (2002). *Underwater acoustic digital signal processing and communication systems* (pp. 37–75). Boston: Kluwer Academic Publishers.
13. Stewart, R. H. (2008). *Introduction to physical oceanography*. Stewart: Robert H.
14. Agarwal, A., Agrawal, M., & Kumar, A. (2019). Higher-order-statistics-based direction-of-arrival estimation of multiple wideband sources with single acoustic vector sensor. *IEEE Journal of Oceanic Engineering, 45*(4), 1439–1449.
15. Singer, A. C., Nelson, J. K., & Kozat, S. S. (2009). Signal processing for underwater acoustic communications. *IEEE Communications Magazine, 47*(1), 90–96.
16. Kaushal, H., & Kaddoum, G. (2016). Underwater optical wireless communication. *IEEE Access, 4*, 1518–1547.
17. Zoksimovski, A., Sexton, D., Stojanovic, M., & Rappaport, C. (2015). Underwater electromagnetic communications using conduction-channel characterization. *Ad Hoc Networks, 34*, 42–51.
18. Siegel, M., & King, R. (1973). Electromagnetic propagation between antennas submerged in the ocean. *IEEE Transactions on Antennas and Propagation, 21*(4), 507–513.
19. Al-Shamma'a, A. I., Shaw, A., & Saman, S. (2004). Propagation of electromagnetic waves at MHz frequencies through seawater. *IEEE Transactions on Antennas and Propagation, 52*(11), 2843–2849.
20. Webster, J. G., & Stojanovic, M. (2015). Underwater acoustic communication. In J. G. Webster (Ed.), *Wiley Encyclopedia of Electrical and Electronics Engineering*.
21. Huang, J. G., Wang, H., He, C. B., Zhang, Q. F., & Jing, L. Y. (2018). Underwater acoustic communication and the general performance evaluation criteria. *Frontiers of Information Technology & Electronic Engineering, 19*(8), 951–971.

22. Ramavath, P. N., Udupi, S. A., & Krishnan, P. (2020). High-speed and reliable underwater wireless optical communication system using multiple-input multiple-output and channel coding techniques for IoUT applications. *Optics Communications, 461,* 125229.
23. Wang, J., Yang, X., Lv, W., Yu, C., Wu, J., Zhao, M., et al. (2019). Underwater wireless optical communication based on multi-pixel photon counter and OFDM modulation. *Optics Communications, 451,* 181–185.
24. Schirripa Spagnolo, G., Cozzella, L., & Leccese, F. (2020). Underwater optical wireless communications: Overview. *Sensors, 20*(8), 2261.
25. Li, Y., Liang, H., Gao, C., Miao, M., & Li, X. (2019). Temporal dispersion compensation for turbid underwater optical wireless communication links. *Optics Communications, 435,* 355–361.
26. Stojanovic, M. (2008). Underwater acoustic communications: Design considerations on the physical layer. In *2008 Fifth Annual Conference on Wireless on Demand Network Systems and Services* (pp. 1–10). IEEE.
27. Godin, O. A. (2008). Sound transmission through water-air interfaces: New insights into an old problem. *Contemporary Physics, 49*(2), 105–123.
28. Wang, X., Qin, T., Qin, Y., Witte, R. S., & Xin, H. (2017). Microwave-induced thermoacoustic communications. *IEEE Transactions on Microwave Theory and Techniques, 65*(9), 3369–3378.
29. Tonolini, F., & Adib, F. (2018). Networking across boundaries: Enabling wireless communication through the water-air interface. In *Proceedings of the 2018 Conference of the ACM Special Interest Group on Data Communication* (pp. 117–131).
30. Long, J., Xie, S., Li, E., Gao, C., Gao, Y., Zhang, Y., et al. (2021). Breakthrough the communication bottleneck between sky and underwater. *AIP Advances, 11*(2), 025029.
31. Fritsch, V. (1963). Propagation of radio frequency electromagnetic fields in geological conductors. *Journal of Research of the National Bureau of Standards-D. Radio Propagation, 67*(2), 161–178.
32. Bedford, M. D., & Kennedy, G. A. (2014). Modeling microwave propagation in natural caves passages. *IEEE Transactions on Antennas and Propagation, 62*(12), 6463–6471.
33. Allum, R. (1986). Cave to surface communications. *Speleonics, 4*(1).
34. Reid, F. (1986). Phone Patch for Handheld Radios. *Speleonics, 4*(4).
35. Abrudan, T. E., Kypris, O., Trigoni, N., & Markham, A. (2016). Impact of rocks and minerals on underground magneto-inductive communication and localization. *IEEE Access, 4,* 3999–4010.
36. Akyildiz, I. F., & Stuntebeck, E. P. (2006). Wireless underground sensor networks: Research challenges. *Ad Hoc Networks, 4*(6), 669–686.
37. Murphy, J. N., & Parkinson, H. E. (1978). Underground mine communications. *Proceedings of the IEEE, 66*(1), 26–50.
38. Einicke, G., Dekker, D., & Hainsworth, D. (1997). A review of underground communications systems. In *Proceedings Technology Exchange Workshop in Coal Mine Productivity.*
39. Patri, A., Nayak, A., & Jayanthu, S. (2013). Wireless communication systems for underground mines-a critical appraisal. *International Journal of Engineering Trends and Technology, 4*(7), 3149–3153.
40. Einicke, G., Dekker, D., & Gladwin, M. (1997). The survivability of underground communication systems following mine emergency incidents. In *Proceedings of the Queensland Mining Industry Health and Safety Conference* (pp. 217–222).
41. Yavuz, A. G., Taysi, Z. C., & Celik, E. (2009). In-cave wireless communication system. In *2009 Third International Conference on Next Generation Mobile Applications, Services and Technologies* (pp. 161–164). IEEE.
42. Ranjan, A., Misra, P., & Sahu, H. B. (2017). Experimental measurements and channel modeling for wireless communication networks in underground mine environments. In *2017 11th European Conference on Antennas and Propagation (EUCAP)* (pp. 1345–1349). IEEE.

43. Defense Advanced Research Projects Agency. (2017). A mechanical based antenna (AMEBA). Retrieved June 14, 2019, from https://www.fbo.gov/spg/ODA/DARPA/CMO/HR001117S0007/listing.html/.
44. Gong, S., Liu, Y., & Liu, Y. (2018). A rotating-magnet based mechanical antenna (RMBMA) for ELF-ULF wireless communication. *Progress In Electromagnetics Research M, 72,* 125–133.
45. Burch, H. C., Garraud, A., Mitchell, M. F., Moore, R. C., & Arnold, D. P. (2018). Experimental generation of ELF radio signals using a rotating magnet. *IEEE Transactions on Antennas and Propagation, 66*(11), 6265–6272.
46. Kemp, M. A., Franzi, M., Haase, A., Jongewaard, E., Whittaker, M. T., Kirkpatrick, M., & Sparr, R. (2019). A high Q piezoelectric resonator as a portable VLF transmitter. *Nature Communications, 10*(1), 1–7.
47. Xu, J., Leung, C. M., Zhuang, X., Li, J., Bhardwaj, S., Volakis, J., & Viehland, D. (2019). A low frequency mechanical transmitter based on magnetoelectric heterostructures operated at their resonance frequency. *Sensors, 19*(4), 853.

Electromagnetic Modeling 3

3.1 Field Generation

The proposed ELF transmitters generate oscillating $\vec{B}$ fields by physically moving one or more static field sources. While in principle any type of periodic motion can be used, in practice mechanical challenges restrict the main choices to oscillations or rotations about a single axis [1–4]. Here we focus on rotating magnetic or electric multipoles, which have the highest possible field generation efficiency [5]. Rotating a high-order multipole enables the frequency of the field pattern to be much larger than that of the motor, which reduces power consumption. However, such multipolar fields decay rapidly with distance: specifically, as $1/r^{(n+1)}$ for the nth order multipole. Since the goal is to project $\vec{B}$ to long distances, we focus on generating and rotating the lowest-order multipoles available, which are dipoles ($n = 2$).

Various motion control methods can be used to implement electromechanical transmitters of this type. For example, spin transfer torques may be useful for implementing highly miniaturized versions based on rotating nanoscale magnetic dipoles. However, in this paper, we limit ourselves to larger devices, for which conventional electric motors are the obvious choice.

The $\vec{B}$ fields generated by rotating magnetic and electric dipoles have been thoroughly analyzed in our earlier work [5]. Here we summarize the main results.

3.1.1 Rotating Magnetic Dipole

Consider a permanent magnet of volume V and remanent flux density B_r magnetized along the z-axis. The resulting dipole moment $\vec{m} = m\hat{z}$ where $m = B_r V/\mu_0$. Now assume that the magnet is rotated at a rate ω about an axis perpendicular to $\hat{z}$ (such as the x-axis) to generate

J. S. Glickstein and S. Mandal, *Electromechanical Transmitters for ELF/VLF Radio*,
Synthesis Lectures on Engineering, Science, and Technology,
https://doi.org/10.1007/978-3-031-15123-1_3

an oscillating $\vec{B}$ field. The resulting dipole moment is circularly polarized and given by $\vec{m(t)} = m\sin(\omega t)\widehat{y} + m\cos(\omega t)\widehat{z}$, i.e., the sum of two oscillating dipoles [5].

The quasistatic approximation is valid in the near-field, i.e., at distances r for which $\omega \ll c/r$, $kr \ll 1$, and $r \ll \lambda/(2\pi)$; here $k = \omega/c$ is the wavenumber and $\lambda = 2\pi/k$ is the EM wavelength. Given the extremely long wavelengths at ELF (at 300 Hz, $\lambda = 10^3$ km in the air), most ELF communication links of interest are entirely in the near-field. Thus, their performance is dominated by the quasistatic components of the generated **B**-field, and the radiated energy is negligible. As a result, the field seen by the receiver is given by

$$\overrightarrow{B_{dip,rot}} = \frac{\mu_0}{4\pi}\frac{m}{r^3}\frac{3}{2}\sum_{l\in\{x,y,z\}}[a_l\cos(\omega t) + b_l\sin(\omega t)]\,\hat{l}, \tag{3.1}$$

where $a_x = \cos(\phi)\sin(2\theta)$, $a_y = \sin(\phi)\sin(2\theta)$, and $a_z = \frac{1}{3} + \cos(2\theta)$, $b_x = \sin(2\phi)$ $\sin^2(\theta)$, $b_y = 2\sin^2(\phi)\sin^2(\theta) - \frac{2}{3}$, $b_z = a_y$, and θ and φ are the elevation and azimuth angles, respectively. Rotating magnetic dipoles behave as relatively omnidirectional sources, as shown in Fig. 3.1a.

For a single-axis receiver coil which is sensitive to a single field component, the maximum available field amplitude $|B_{dip,rot}|$ is

$$B_{dip,max} = 2\frac{B_r V}{4\pi r^3}. \tag{3.2}$$

All three field components oscillate in phase, so they can also be easily summed up by a multi-axis receiver coil to generate the total field amplitude $|B_x + B_y + B_z|$, which has a maximum value given by

$$B_{dip,max} = \frac{3\sqrt{3}+1}{2}\frac{B_r V}{4\pi r^3}, \tag{3.3}$$

which occurs when $\theta = 0$ or π. This is 55% larger than for the single-axis case.

3.1.2 Rotating Electric Dipole

An electric dipole p_0 of volume V can also be rotated about an axis perpendicular to itself to generate an oscillating $\vec{B}$ field; typical values of remanent polarization $P_r = p_0/V$ for ferroelectric or piezoelectric materials are 20–50 μ C/cm^2. The field due to the time-varying polarization vector $\widehat{p} = p_0[\sin(\omega t)\widehat{y} + \cos(\omega t)\widehat{z}]$ is

$$\overrightarrow{B_{ed,rot}} = \frac{\mu_0}{4\pi}p_0\omega(1 - ikr)\frac{e^{ikr}}{r^2}\cdot\sum_{l\in\{x,y,z\}}[a_l\cos(\omega t) + b_l\sin(\omega t)]\hat{l}, \tag{3.4}$$

where $a_x = -\cos(\theta)$, $a_y = 0$, $a_z = \cos(\phi)\sin(\theta)$, $b_x = \sin(\phi)\sin(\theta)$, $b_y = -\cos(\phi)$ $\sin(\theta)$, and $b_z = 0$. This function is the sum of a near-field term $\propto \omega/r^2$ which dominates when $kr \ll 1$ and a far-field term $\propto \omega^2/r$ which dominates when $kr \gg 1$. Thus, rotating electric dipoles also behave as relatively omnidirectional sources. However, the spatial distribution of total field amplitude $|B_x + B_y + B_z|$ is significantly different from the rotating magnetic dipole (see Fig. 3.1b). It has a maximum of

$$|\vec{B}_{ed,rot}| = \sqrt{3}\,(\mu_0\omega)\;P_r V/\left(4\pi r^2\right) \tag{3.5}$$

in the near-field or

$$|\vec{B}_{ed,rot}| = \sqrt{3}\left(\mu_0\omega^2/c\right)\;P_r V/\left(4\pi r\right) \tag{3.6}$$

in the far-field.

3.1.3 Effect of Ground Planes

At ELF frequencies with wavelengths beyond millions of meters, the Earth is an effective ground plane for communication from any reasonable elevation of the transmitter above mean sea level [6]. The strength of the ELF $\vec{B}$ field produced at the location of the receiver by a rotating dipole in the air is effectively doubled by the presence of the Earth. Therefore, the model developed in Eq. (3.1) for a dipole in free space predicts half the $B_{dip,rot}$ that is observed in practice as a result of the ground plane effect.

3.1.4 Comparison Between Magnetic and Electric Dipoles

$\vec{B}$-field patterns for rotating magnetic and electric dipoles are shown in Fig. 3.1. While the two patterns are broadly similar, the near-field generated by a magnetic dipole decreases with distance as r^{-3}, while that from an electric dipole decreases more slowly (as r^{-2} and r^{-1} in the near- and far-field, respectively). Also the former is frequency-independent, while the latter increases with frequency as ω and ω^2, respectively. Thus, electric dipoles are preferable for longer-range and higher-frequency applications.

3.2 Power Consumption

The peak electrical power input of a mechanical source is

$$P_{elec} = \frac{(\tau_f + \tau_{mod})\,\omega}{\eta_m \eta_e} = \left(\tau_f + J\frac{d\omega}{dt}\right)\frac{\omega}{\eta_m \eta_e}. \tag{3.7}$$

Here J is the total moment of inertia of all moving parts (magnet, bearings, and rotor), τ_f is the torque due to mechanical losses (air friction, bearing loss, etc.), and $\tau_{mod} = J(\omega t)$ is the torque due to frequency modulation (used to transmit data). Also, η_m and η_e are the efficiency of the motor and the drive electronics, respectively. The components of τ_f due to

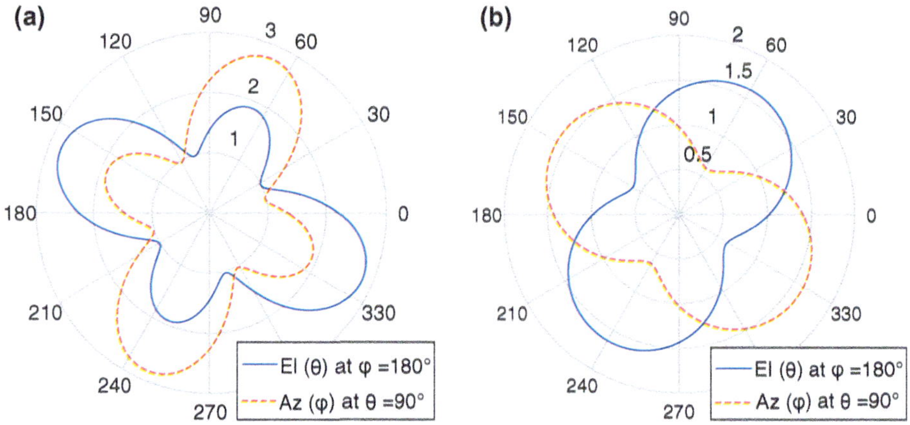

Fig. 3.1 Theoretical near-field **B**-field patterns created by (a) a rotating magnetic dipol, and (b) a rotating electric dipole. In each case, elevation (El) and azimuth (Az) patterns of $|B_x + B_y + B_z|$ (the total amplitude measured by a 3-axis **B**-field sensor) are shown, where θ and ϕ denote the elevation and azimuth angles, respectively

air friction and bearing loss scale as $\omega^2 V^2$ and ωV, respectively, where V is the volume of the source [7]. Thus, the mechanical power consumption $\omega\tau_f$ required to generate a target $\vec{B}$-field magnitude increases rapidly with ω, although the increase is less rapid for rotating electric dipoles that generate stronger $\vec{B}$-fields as ω increases, which allows V to be reduced while keeping $|B|$ constant.

Figure 3.2 summarizes this analysis by showing qualitative estimates of the power required to generate a given $\vec{B}$-field amplitude using rotating magnetic and electric dipoles as a function of frequency (assuming the total source volume remains fixed). The power consumption of a conventional coil-based near-field transmitter is also shown for comparison. Note that the plot shows the *normalized* power consumption of these sources. Rotating a magnet (near-field), electret (near-field), or electret (far-field) at ω results in a mechanical power consumption P that scales as ω^3, ω^2, and ω, respectively. On the other hand, using oscillating current in a loop as a magnetic dipole results in electrical power consumption P that scales as $(1 + c\omega^{1/2})$ where c is a constant; here the first term is due to DC resistance and the second due to skin effect. Thus, mechanical transmitters are beneficial (i.e., have the lowest power consumption for a given field strength) up to a certain maximum operating frequency f_{max}. Next, we discuss design methods to reduce the power consumption of all-mechanical transmitters, thus increasing the value of f_{max}.

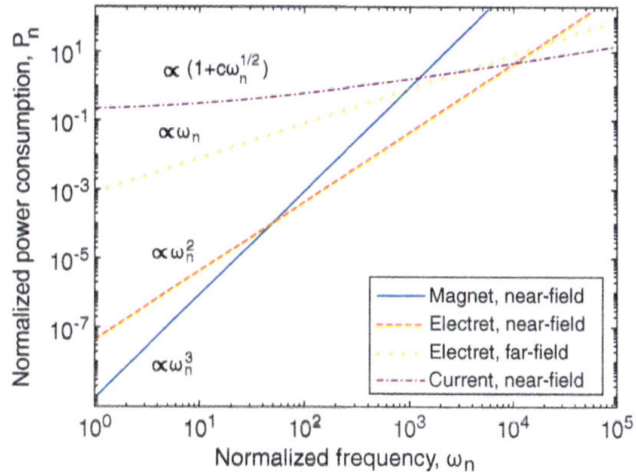

Fig. 3.2 Qualitative comparison of the power consumed by different sources in order to generate the same **B**-field intensity at a given distance from an ELF transmitter

3.2.1 Shape of the Dipole

The shape of the rotating source (magnet or electret) does not affect the dipole field, which only depends on volume V. Thus, we can reduce power consumption by minimizing J while keeping V fixed to maintain field strength. For example, axially rotating cylinders of length l have $J = \rho V^2/(2il)$, so J can be reduced by increasing the aspect ratio l/R of the cylinder while keeping V constant. Moreover, large aspect ratios also tend to reduce air friction. Thus, we focus on diametrically polarized and axially rotating cylinders with large aspect ratios as field sources. However, the resulting long and thin shapes are fragile and prone to mechanical instability during high-speed rotation (e.g., due to the advent of lateral oscillations) [8, 9].

3.2.2 Distributed Transmitters

The strong dependence of inertia on physical size (e.g., $J \propto R^4$ for cylinders) suggests that *distributed transmitters* consisting of an array of small sources should be more power-efficient than a single large source of the same total volume. For example, consider dividing the source into $N > 1$ identical synchronously rotating elements that are scaled down as $R \propto N^{-1/2}$ to keep the overall volume V constant. In this case, J per array element decreases as $1/N^2$.

Air friction is a major contributor to τ_f. We assume that each dipole is tightly packed inside a hollow cylindrical shaft (thickness t, length $l \gg R$) to improve mechanical stiffness and that the latter rotates inside a larger cylindrical chamber of radius $(R_s + \delta)$, where $R_s \equiv (R + t)$ is the outer radius of the shaft and δ is the air gap. Air friction loss in this geometry has been analyzed and experimentally validated [7]. The resulting torque for each shaft is

$$\tau_f = c_f i \rho_{air} \omega^2 R_s^4 l \tag{3.8}$$

521545_1_En_3_Chapter-print ☑ TYPESET ☐ DISK ☐ LE ☑ CP Disp.:9/10/2022 Pages: xxx Layout: German_T5

where c_f is the friction coefficient and ρ_{air} is the air density. The value of c_f varies with geometry and speed: (i) $c_f \propto 1/Re$ for laminar flow, where $Re = R_s^2\omega/\nu$ is the Reynolds number and ν is the kinematic viscosity of air; and (ii) $c_f \propto Ta^{-1/5}$ once vortices form, where $Ta = (R_s\omega\delta/\nu)\sqrt{\delta/R_s}$ is the Taylor number. The transition between (i) and (ii) occurs at $Ta \approx 41.3$. Overall, we see that the torque due to air friction decreases as $1/N^2$.

3.2.3 Dynamic Stability

The bending modes of the rotating shaft should be avoided to (i) minimize mechanical failures due to *whirling* (transverse vibrations) and (ii) reduce acoustic noise [9]. Ideally, the critical speeds corresponding to these resonances should be higher than the desired rotation speed. The first critical speed can be approximately computed by the Rayleigh-Ritz (R-R) method [10] to be $\omega_{cr} \approx \sqrt{g/y_{max}}$, where g is the acceleration due to gravity and y_{max} is the maximum static deflection of the shaft. Evaluating y_{max} results in

$$\omega_{cr} \approx \alpha\sqrt{\frac{EI}{Ml^3}}, \tag{3.9}$$

where E is Young's modulus, $I = iR^4/4$ is the second moment of area, and α is a dimensionless constant that depends on the loading conditions of the shaft (e.g., a supported beam or a clamped beam) [11].

The analysis above shows that ω_{cr} rapidly decreases as the shaft gets longer. The risk of operating above ω_{cr} can be mitigated using two approaches: (i) developing magnet assemblies with higher remanent magnetization B_r to reduce the magnet volume V and (ii) developing motor control methods to safely accelerate the shaft through ω_{cr}. Alternatively, ω_{cr} can be increased by encasing the magnets within a thin "shell" of low density and high stiffness (e.g., made of a titanium or beryllium alloy). The R-R method and superposition can then be used to estimate the contributions of the two rotating components (magnet and shell) to the effective stiffness of the composite shaft. In particular, assuming the rotor to be a point mass, the two critical speeds can be combined using Dunkerley's formula [12], which predicts that $\omega_{cr}^{-2} = \sum_i \omega_{cr,i}^{-2}$. This result can be refined using finite element analysis.

The total moment of inertia (including both the dipoles and shafts) is

$$J = N\left(MR^2 + M_s\left(R_s^2 + R^2\right)\right)/2. \tag{3.10}$$

This can be written as

$$J(N) \approx \rho V^2/(2\pi lN) + 2\rho_s V^{3/2} t/\sqrt{\pi lN} \tag{3.11}$$

when $t \ll R$, which is a decreasing function of N; note that M_s and ρ_s are the mass and density of the shaft, respectively. Similarly, the total air friction torque $N\tau_f$ also decreases with N because of the strong dependence of τ_f on shaft radius. Bearing loss, which is included within the efficiency η_m, will also behave similarly with N. Thus, the peak electrical power $P_{elec}(N)$ decreases significantly with N.

To summarize, J per array element decreases as $(V/N)^2$ and torque due to air friction decreases as $1/N^2$. The generated field per array element decreases as V/N, and thus overall field magnitude remains constant while the net power consumption $P_{elec}(N)$ decreases for the system of N elements as N increases. Thus, the effects of mechanical inertia on the system are reduced: *it is more power-efficient to drive an array of small rotating dipoles.* Moreover, such *distributed transmitters* also provide built-in redundancy and robustness to localized failures. It is worth noting that we do not consider such distributed transmitters to be useful as phased arrays. This is because the very long wavelengths at ELF (e.g., 10^5 m at 3 kHz in the air) ensure that even km-scale arrays have sub-wavelength apertures. As a result, the fields from individual sources simply add in phase.

3.2.4 Interactions Between Rotating Dipoles

One challenge with using distributed transmitters is the dipole-dipole coupling between nearby sources, which increases power consumption by creating an additional *coupling torque* τ_c that must be supplied by the motors. For example, in the prototypical case of identical magnets of moment m separated by a distance $d = d\hat{z}$ and rotating synchronously about the x-axis, the coupling torque is

$$\vec{*\tau C} = \vec{m} \times \vec{B} = \frac{3\mu_0 m^2}{8\pi d^3} \sin(2\omega t)\hat{x}, \tag{3.12}$$

which fluctuates sinusoidally at a frequency 2ω while decreasing with distance as $1/d^3$. Coupled electrets with a dipole moment p behave similarly: the resulting torque is the same as Eq. (3.12) once $\mu_0 m^2$ is replaced with p^2/ϵ_0.

While the analysis above predicts that τ_c oscillates at a frequency of 2ω in both cases, in practice the fluctuations in τ_c are random since ω is modulated to transmit data [5]. Fortunately, much of the electrical power required to supply such quasi-oscillatory torques can be recovered by using regenerative braking methods [13]. Nevertheless, the motor drivers must be appropriately sized to handle the rapidly fluctuating peak power requirements.

The dipoles should be positioned within the available system volume in a way that minimizes the total coupling torque and interaction energy. In 1D, the minimum energy configuration consists of uniformly spaced dipoles rotating in alternating directions. In 2D and 3D, a general solution that minimizes energy for arbitrary N is not known. However, we can refer to the analogous Thomson problem in electrostatics [14]. Specifically, N electric charges confined to a thin circular disk will space themselves equally around the perimeter for $N \leq 11$, while some of them will reside in the interior for $N > 11$. By contrast, charges confined to a spherical region will distribute themselves on its surface for any value of N. There is no general analytical solution for this distribution, although specific ones are

521545_1_En_3_Chapter-print ☑ TYPESET ☐ DISK ☐ LE ☑ CP Disp.:9/10/2022 Pages: xxx Layout: German_T5

known: for $N = 2, 4, 6$, and 12, the charges form the vertices of a regular digon, tetrahedron, octahedron, and icosahedron, respectively. These solutions can be taken as starting points for torque calculations. In practice, it is often sufficient to only consider interactions between nearest neighbors; this is a good approximation because the torques decay rapidly with distance, i.e., as ($\tau_c \propto d^{-3}$).

References

1. Gong, S., Liu, Y., & Liu, Y. (2018). A rotating-magnet based mechanical antenna (RMBMA) for ELF-ULF wireless communication. *Progress In Electromagnetics Research M, 72*, 125–133.
2. Burch, H. C., Garraud, A., Mitchell, M. F., Moore, R. C., & Arnold, D. P. (2018). Experimental generation of ELF radio signals using a rotating magnet. *IEEE Transactions on Antennas and Propagation, 66*(11), 6265–6272.
3. Kemp, M. A., Franzi, M., Haase, A., Jongewaard, E., Whittaker, M. T., Kirkpatrick, M., & Sparr, R. (2019). A high Q piezoelectric resonator as a portable VLF transmitter. *Nature Communications, 10*(1), 1–7.
4. Xu, J., Leung, C. M., Zhuang, X., Li, J., Bhardwaj, S., Volakis, J., & Viehland, D. (2019). A low frequency mechanical transmitter based on magnetoelectric heterostructures operated at their resonance frequency. *Sensors, 19*(4), 853.
5. Madanayake, A., Choi, S., Tarek, M., Dharmasena, S., Mandal, S., Glickstein, J., et al. (2017). Energy-efficient ULF/VLF transmitters based on mechanically-rotating dipoles. In *2017 Moratuwa Engineering Research Conference (MERCon)* (pp. 230–235). IEEE.
6. Burch, H. C., Garraud, A., Mitchell, M. F., Moore, R. C., & Arnold, D. P. (2018). Experimental generation of ELF radio signals using a rotating magnet. *IEEE Transactions on Antennas and Propagation, 66*(11), 6265–6272.
7. Zwyssig, C., Round, S. D., & Kolar, J. W. (2006). Analytical and experimental investigation of a low torque, ultra-high speed drive system. In *Conference Record of the 2006 IEEE Industry Applications Conference Forty-First IAS Annual Meeting* (Vol. 3, pp. 1507-1513). IEEE.
8. Jeffcott, H. H. (1919). XXVII. The lateral vibration of loaded shafts in the neighbourhood of a whirling speed.– The effect of want of balance. *The London, Edinburgh, and Dublin Philosophical Magazine and Journal of Science, 37*(219), 304–314.
9. Swanson, E., Powell, C. D., & Weissman, S. (2005). A practical review of rotating machinery critical speeds and modes. *Sound and Vibration, 39*(5), 16–17.
10. Trefethen, L. N., & Bau III, D. (1997). *Numerical linear algebra* (Vol. 50). SIAM Publications.
11. Tenconi, A., Vaschetto, S., & Vigliani, A. (2013). Electrical machines for high-speed applications: Design considerations and tradeoffs. *IEEE Transactions on Industrial Electronics, 61*(6), 3022–3029.
12. Atzori, B. (1974). Dunkerley's formula for finding the lowest frequency of vibration of elastic systems. *Journal of Sound and Vibration, 36*(4), 563–564.
13. Dixon, J. W., & Ortuzar, M. E. (2002). Ultracapacitors+ DC-DC converters in regenerative braking system. *IEEE Aerospace and Electronic Systems Magazine, 17*(8), 16–21.
14. Calkin, M. G., Kiang, D., & Tindall, D. A. (1987). Minimum-energy charge configurations. *American Journal of Physics, 55*(2), 157–158.

521545_1_En_3_Chapter-print ☑ TYPESET ☐ DISK ☐ LE ☑ CP Disp.:9/10/2022 Pages: xxx Layout: German_T5

Design of ELF Communications Links

4

4.1 Natural and Artificial Activity in the ELF Channel

ELF channels are home to many natural and artificial signals. The brain and heart of higher life forms produce both electric and magnetic fields in the ELF range [1]. Perception processing in the visual cortex of humans is associated with activity in the 30–90 Hz range, while high-frequency 200 Hz signals are associated with memory storage in the neocortex during sleep [2]. Geological activities produce a diverse range of uniquely characterizable ELF and VLF emissions. In many cases, the root source of the effect is lightning, though associated with various activities such as thunderstorms, volcanic eruptions, dust storms, and so on [3].

Weather on distant planets radiates EM energy at ELF/VLF frequencies similar to our own planet. In the prior art, the atmospheres of other planets in the solar system have been studied through analysis of electromagnetic field measurements acquired by space probes passing by (including Voyager 1 and Voyager 2) and descending into (including Venera) distant atmospheres [4]. Electric fields in the Van Allen radiation belt of our planet show activity at 10–100 kHz which is studied with VLF radio space probes [5].

Man-made machinations are also efficient producers of ELF/VLF signals, though unfortunately not in any way which is conducive to ELF communications. Power lines radiate EM fields at 50 Hz and 60 Hz, transport railways in some regions of Europe operate at $16\frac{2}{3}$ Hz, and higher harmonics from these signals are present across the spectrum as well.

Many natural sources of ELF signals have been correlated to changes in biological functions, which are the focus of extensive scientific research. While none have identified links between ELF and negative health effects, public perception of RF radiation in general [6] has strong implications in the construction and management of enormous ELF transmitter stations.

J. S. Glickstein and S. Mandal, *Electromechanical Transmitters for ELF/VLF Radio*,
Synthesis Lectures on Engineering, Science, and Technology,
https://doi.org/10.1007/978-3-031-15123-1_4

4.1.1 Atmospheric Effects

Sferics, electromagnetic impulses emitted by lightning, are contained mostly within the ELF/VLF band [7]. Within the ionosphere, there is little attenuation (~2–3 dB/1000 km, [8]) and signals can travel thousands of kilometers before detection. Broadband sferics observed during a lightning storm are shown in Fig 4.1.

The phenomenon is useful in atmospheric monitoring and detection of lightning strikes. Over long distances, propagation of sferics shows dispersion effects resulting from the varied penetration of the ionosphere. Dispersion effects are heavily influenced by local atmospheric conditions near the source and ionospheric conditions across the channel. Weather patterns in various seasons and across temperature and humidity gradients show pronounced effects in the propagation of ELF signals through the ionosphere [9, 10]. Observation of these sferics is useful in monitoring global weather patterns [11, 12].

Some sources describe these sferics distinctly as "tweeks" when their origin is the ionosphere. Such signals have slight time dispersion effects as may be observed in Fig. 4.2. VLF radio engineers observe the time distortion of these signals to analyze the conditions of a communication channel between two points on the globe [11]. A large amount of research and publication occurred during the 1960s and 1970s when ELF/VLF radio was first studied and many of the stations in present use were constructed.

EM radiation (preferentially at low frequencies) interactions in the magnetosphere, when some EM radiation penetrates the lower barrier of the ionosphere and enters the Van Allen radiation belts, cause greater time dispersion than tweeks. The dramatic change in prop-

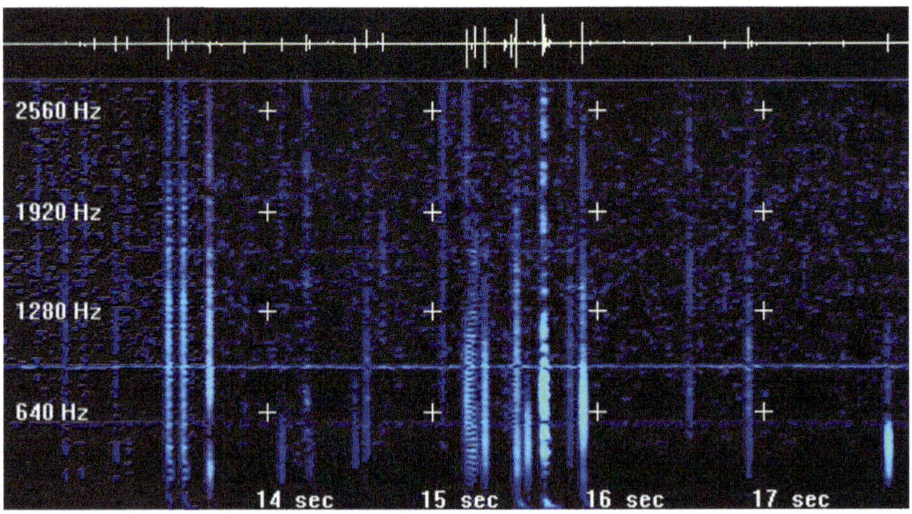

Fig. 4.1 Sferics captured by VLF radio at NASA's Marshall Space Flight Center

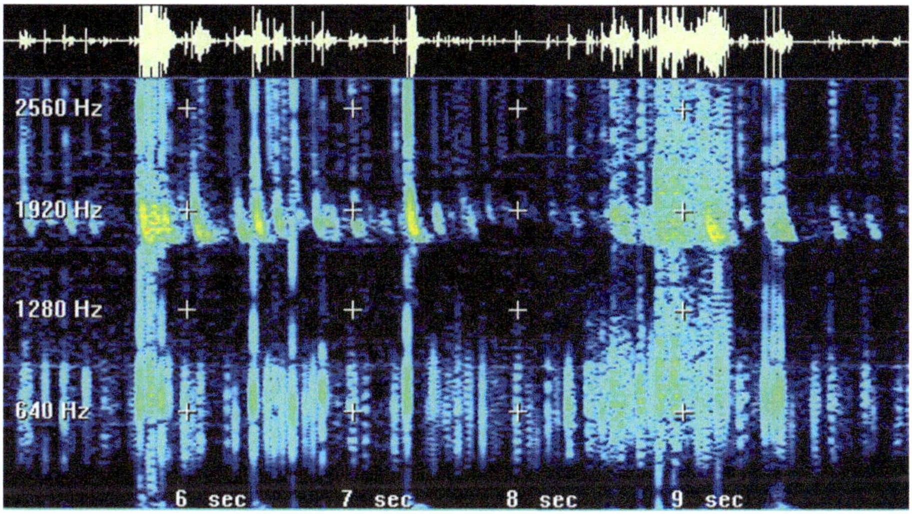

Fig. 4.2 ELF/VLF tweeks showing time-dependent frequency dispersion

agation path delay (preferentially at low frequencies) due to far lower attenuation in the magnetosphere produces *whistlers*. These whistlers have distinct spectrograms such as that of Fig. 4.3.

4.1.2 Schumann Resonance

The global Earth-ionosphere cavity forms a closed waveguide and exhibits resonances. This behavior was first discovered by W. O. Schumann, who demonstrated that the resonance frequencies are given by $f_n = 7.49(n(n+1))$. The fundamental ($n = 1$) resonant mode occurs at 10.6 Hz with higher tones at 18.4, 26.0, 33.5, and 41.1 Hz. Project Sanguine (and the later Project ELF, see Sect. 2.1.1) transmitters operated between the ninth and tenth Schumann resonances. It is reported that the intention was to use ionospheric resonance to increase transmitter range [3].

Cosmic activities including weather on other planets in the solar system and gamma-ray bursts in distant galaxies excite ionospheric resonance modes which have been identified and studied [13]. Other sources of extra-terrestrial energy emission and their prevalence in the background noise observable in the ionosphere are of recent interest in the field.

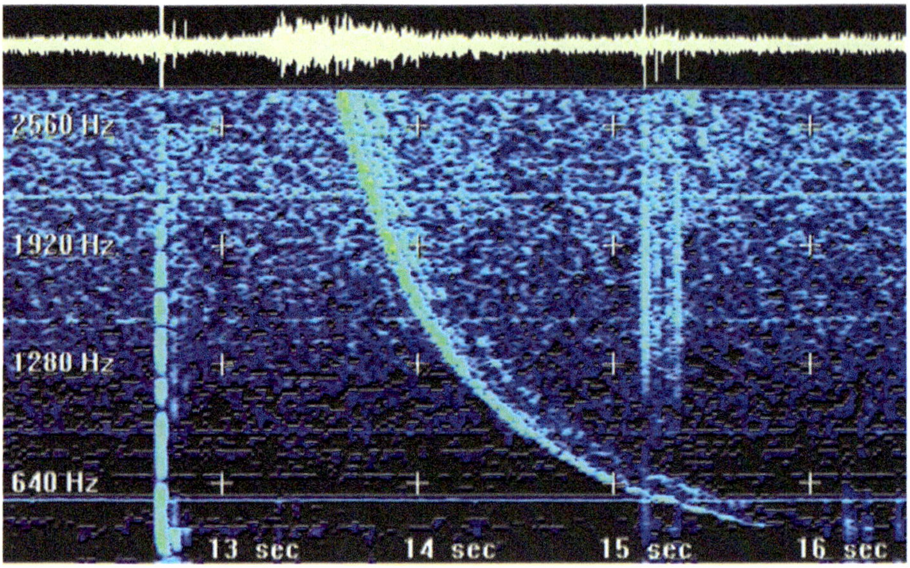

Fig. 4.3 Whistlers have frequency dispersion on a scale of seconds

4.1.3 Animal Response

The potential effects of electric and magnetic ELF fields on biological systems have been studied extensively since the dawn of ELF radio. Data indicate that the presence of ELF fields does not significantly affect *behaviors* in any consistent manner, though some physiological responses have been noted which direct further investigation [1, pp. 275–276, 295]. However, many correlations are documented between natural and artificial ELF fields and changes in biological *function*, which have generated a lot of attention [14]. Biochemical reactions, metabolic function, and even emotional responses are reportedly associated with various low-frequency EM fields over a short or long duration. Metabolic rates vary in goldfish and frogs. Human blood pressure appears to show a correlation with the intensity of Schumann resonance [15]. Occupational studies report that exposure to ELF fields between 16.7–60 Hz significantly reduces melatonin levels [16].

In some cases, these biological responses have not been shown to present danger or interrupt physiological/cognitive function; while research continues to investigate all possible negative health effects, the current science is focused on searching for explanations to natural functions. It is suggested, for example, that animal activity before thunderstorms is driven by sensory perception of changing ELF fields. Neurological functions such as sleep are associated with effects from cosmic EM fields [17]. The literature suggests it might be possible for EM fluctuations to affect human physiology, psychology, and behavior in ways we have not yet identified, since the magnetoreception of neural structures should be evolutionarily adjusted to ambient magnetic fields [15]. Some research has even proposed models for ELF field effects enabling the evolution of life from inanimate substances (abiogenesis) [1]. It is difficult to study these effects as many background fields persist in nature, such that shielding subjects from all but one source and producing a controlled environment remains a barrier.

In the absence of a known and credible biophysical mechanism to explain such effects, the idea that natural or man-made ELF fields might influence human health on earth appears to lack scientific credibility. Nevertheless, significant mutual correlations observed between hospital admissions, solar-geomagnetic activity, and ionospheric ion/electron concentrations continue to fuel speculation in both the general public and the academic community. In [16], Cherry describes several hypotheses proposed in the prior art to explain how cause-and-effect relationships to serious human health effects could exist. For example, perhaps exposure to ELF fields reduces melatonin levels and this effect causes some illness due to decreased antioxidant activity in the body. If well-supported biophysical mechanisms to explain these processes are found to exist in future work, the discovery would result in a major paradigm shift in human knowledge. Rosen et al. [18] present evidence that exposure to a 50 μT 60 Hz ELF field increases serotonin and reduces melatonin in laboratory mice. Reiter and Robinson [19] report reduced melatonin is associated with neurological, cardiac, reproductive, and carcinogenic illness. Without an explanation for these alluring correlations, it is easy to see how speculation over ELF fields has continued for so many decades. The pervasive correlation between ELF fields and human health remains a peculiar phenomenon with unknown significance.

While research has yet to demonstrate a direct link between EM fields and adverse health effects, social factors weigh heavily on public perception. The question of whether mankind or other organisms are sensitive to any significant extent is one that has been raised as a concern in the public eye since the construction of the first ELF transmitters began in the 1960s. In the absence of evidence supporting adverse health effects, documented biological changes associated with ELF fields are enough to cause some individuals to feel psychological effects with placebo indications, or "nocebo effects" [20, 21]. Unlike ELF fields, this nocebo effect has been documented to produce adverse health effects leading some to experience self-reported "ELF allergies" in response to the mere suggestion of exposure to low-frequency EM fields [22].

4.2 Characterization of the ELF Channel

The actual performance of the transmitter in an established communication link is heavily dependent on the conditions of the channel, which are location-dependent. While some general characteristics were described in the previous section, additional location-specific information is required for setting up an ELF communications link. Accordingly, a field survey was conducted to study the power spectral density of noise in the ELF range across

Fig. 4.4 Two locations are marked on this map of the Cuyahoga Valley National Park where ELF noise spectra are collected

central and northern Ohio, USA, in which the same receiver configuration as in the prototype was used to collect data.

The survey began in the Cuyahoga Valley National Park, where the goal was to make observations as far away from power-line interference as possible. The U.S. Energy Information Administration (EIA) publishes data on the distribution of the power grid, including information they collect from power distribution companies [23]. Using their data, we sketched the position of the major power lines on a map of the park, and the park rangers were kind enough to point the authors toward potential areas to survey. The Oak Grove Picnic Area market in Fig. 4.4 is the oldest area of the park and most likely for park administrators to leave untouched. Data was also collected at the visitor center located ~6.25 km (~3.75 mi) southwest of there.

The power spectral density (PSD) of Fig. 4.5 was measured at the visitor center with 10 s acquisition at 100 kSa/s. The bandwidth of interest is at most 1 kHz for AMEBA; oversampling by 50× the Nyquist rate reduces instrumentation noise by ~7×. Acquisition time at the location is far longer (100 s) than what is used for the presented figures, which are produced with intentionally truncated sections of data to maintain a step size on the x-axis which is appropriate for accurate portrayal in print.

The measurement was repeated in a remote area of the park and the result is shown in Fig. 4.6. Observe that 60 Hz harmonics are present in both cases though the power of higher harmonics is significantly lower in the remote area. Power-line interference (most notably at lower harmonics) has a peculiar spectral spread which is present in addition to distinct spikes at $60 \times N$ Hz. The effect varies at different locations and it appears the characteristic is not necessarily correlated to peak values (though it might suggest indicators for the types of power generation and usage present nearby each location).

The 2–200 Hz ELF band is maximally interesting, given the typical rotational speeds of commercially-available electric motors. Observing Fig. 4.7, the greatest noise content

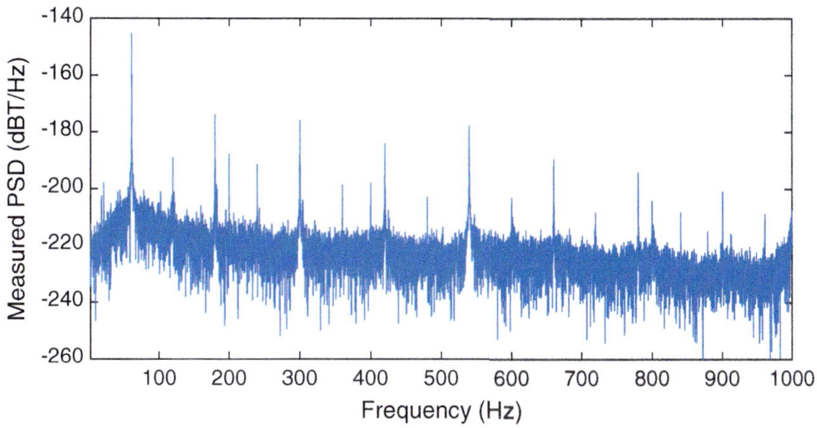

Fig. 4.5 Magnetic field power spectral density measured at the visitor center of Cuyahoga Valley National Park

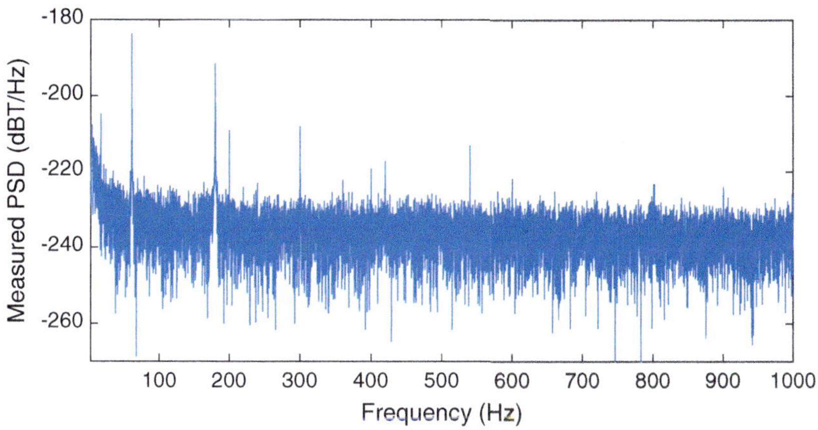

Fig. 4.6 Magnetic field power spectral density measured near the Oak Grove picnic area of Cuyahoga Valley National Park

in this band is present in University Heights, OH, and comparable noise is present in a suburb of Parma, OH. However, the noise in a residential suburb of New Albany, OH, is significantly lower. Possible influences are the number of electricity users, the density of nearby commercial shopping and industrial complexes, the age of the electrical grid, and the quality of local power distribution infrastructure. University Heights and Parma have fairly dense residential and commercial districts, while New Albany is mostly farmland and was a village some 10 years ago.

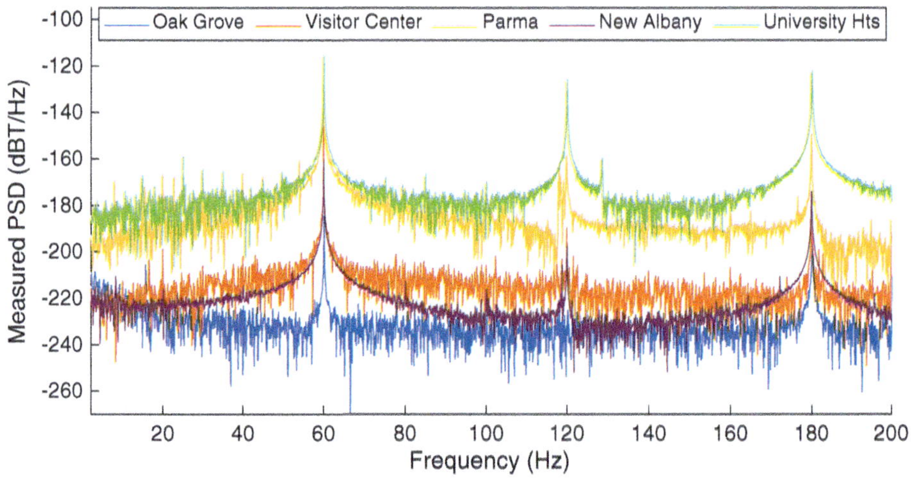

Fig. 4.7 Magnetic field PSD at five locations across central and northern Ohio, USA, presented in direct comparison

4.3 System Design

A block diagram of a single-element electromechanical ELF transmitter is shown in Fig. 4.8; distributed designs are considered later. Most of the design complexity resides within the digital controller, which (i) performs power management functions, (ii) generates the modulation waveform given the input data, and (ii) controls the motion of the field source (assumed to be a cylindrical permanent magnet) through a DC-AC power converter (inverter) and motor driver.

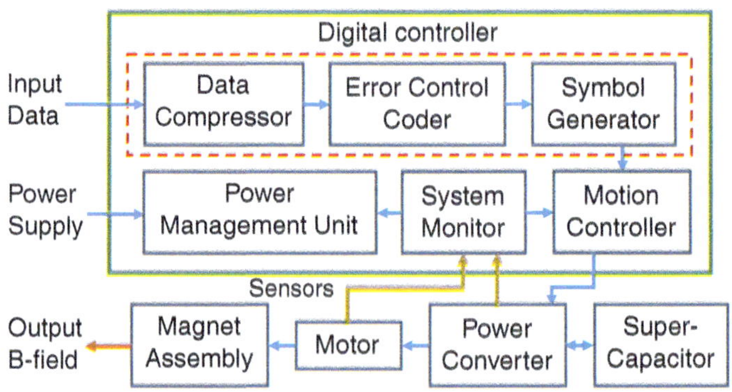

Fig. 4.8 Proposed system design for transmitter producing $\vec{B}$-field signals from supplied input data

4.3.1 Data Modulation

The data modulation scheme, which is shown within the dashed box in Fig. 4.8, is the focus of much concern in the design of power-efficient ELF communications. The proposed scheme should address the fundamental problem of power-efficient control of a mechanical device with high inertia and stored energy. We define efficient modulation schemes as those which minimize the additional torque τ_{mod} required by the motor to maintain the desired bit rate R_B. Given R_B, we then wish to minimize the symbol rate R_S by maximizing the number of bits within each symbol. Continuous phase modulation (CPM) methods such as continuous phase frequency-shift keying (CP-FSK), Gaussian minimum shift keying (GMSK), and shaped-offset quadrature phase-shift keying (SOQPSK) are promising for this purpose. The resulting waveforms have (i) a continuous phase, which results in high spectral efficiency and low out-of-band emissions, and (ii) a constant envelope (CE), which maximizes power efficiency [24]. In addition, rotating dipoles naturally generate CE waveforms; therefore, CPM is well matched to the physical constraints of electromechanical transmitters. However, the initial phase of each CPM symbol depends on the earlier ones (i.e., the symbols have phase memory), which makes the design and implementation of an optimal receiver more complicated. The optimum receiver operates by detecting sequences of symbols, rather than on a bit-by-bit basis [25]. Fortunately, the necessary methods for detecting sequences, such as maximum likelihood sequence estimation (MLSE) using the Viterbi algorithm, are well known and can be readily implemented in real time at the low data rates (<1 kbps) supported by ELF links.

4.3.2 Minimizing Power Consumption

The proposed approach to enable power-efficient and highly miniaturized wireless ELF data transmission in our experimental prototype consists of two steps. The first is source coding to compress the data by removing redundancy. The second is to minimize the modulation torque τ_{mod} and associated power $\omega\tau_{mod}$ supplied by the motor while maintaining the desired bit rate R_B.

4.3.2.1 Data compression

The total energy required to transmit a given message within a specified time can be reduced by compressing the source data, which in turn reduces the necessary bit rate R_B. For example, consider DEFLATE, which is a well-known lossless data compression algorithm used in PNG images, ZIP files, etc. [26]. DEFLATE is a computationally efficient scheme that combines the Lempel-Ziv algorithm (LZ77) [27] with Huffman coding to compress typical plain text by $\approx 3\times$. However, DEFLATE operates on data encoded in ASCII, which requires 8 binary bits per character. To further compress the input message, a reduced character set is implemented consisting of numbers 0–9, letters A–Z, and a limited set of punctuation (space, comma, period, semicolon, and exclamation mark). Using this reduced set of at most 49 characters, only 6 binary bits are required per character. Nevertheless, DEFLATE is written to operate on ASCII data; thus, we expect further improvements in compression ratio by using specialized compression algorithms that make effective use of the proposed reduced character set. Such a scheme is described in Sect. 4.3.4.

4.3.2.2 Modulation Waveform

Energy consumption can be further reduced by packing multiple bits within each symbol to decrease the symbol rate R_S. In earlier work, a multiple frequency-shift keying (MFSK)-based scheme has been developed for this purpose [28]. MFSK is a M-ary orthogonal modulation scheme in which M is the size of the alphabet (number of frequencies) and each symbol represents $\log_2(M)$ bits, resulting in $R_S = R_B/\log_2(M)$. However, a standard MFSK waveform has discontinuous phase $\varphi(t)$ due to rapid jumps between the symbols, which results in large spikes in the instantaneous frequency $\omega(t) = d\varphi(t)/dt$. CP-FSK is a well-known variant of MFSK that keeps $\varphi(t)$ continuous, i.e., eliminates frequency spikes by generating a CPM waveform [29].

A generic M-ary CPM waveform $s(t)$ is defined as

$$s(t) = A_c \cos(\omega_0 t + 2\pi \sum_{i=0}^{n} \alpha_i h_i q(t - iT)), \tag{4.1}$$

where amplitude A_c is a constant proportional to the dipole moment; $T = \frac{1}{R_S}$ is the symbol period; α_i are the $\log_2(M)$-bit data symbols; h_i is the modulation index (which can vary from symbol to symbol in multi-h schemes); and $q(t)$ is the *phase response function*. CPM schemes can be classified using $q(t)$ or its derivative $dq/dt = g(t)$, which is known as the *frequency response function*. For example, LRC and $LREC$ schemes use raised cosine and rectangular frequency pulses of length L symbols, respectively. The term CP-FSK refers to $LREC$ schemes with $L = 1$ that define $g(t)$ to be $1/(2T)$ for $0 < t < T$ and zero otherwise. Thus, they are commonly referred to as $1REC$, denoting CPM schemes with rectangular frequency pulses of length T and phase memory $g(t)$ communicated across one symbol. CP-FSK waveforms are continuous, but the symbols are not orthogonal unless h is a multiple of $\frac{1}{2}$.

In order to understand the properties of CP-FSK, it is useful to first consider a conventional MFSK waveform $s(t)$, which is defined as

$$s(t) = A_c \cos(\omega_0 t + 2\pi \Delta f \times m(t)). \tag{4.2}$$

Here $m(t)$ is the digital data to be transmitted, Δf sets the frequency steps between symbols, and ω_0 is an integer multiple of Δf to ensure orthogonality between them. Signals for which

$m(t) \in [0, 1]$ result in binary FSK, where each symbol contains a single bit. Similarly, signals for which $m(t) \in [0, M - 1]$ result in MFSK.

By contrast, a M-ary CP-FSK waveform is defined as

$$s(t) = A_c \cos\left(\omega_0 t + 2\pi \Delta f \int_0^t m(t_1) dt_1\right), \tag{4.3}$$

where all terms remain as previously defined. Comparing Eqs. (4.1) with (4.2), we see that the instantaneous phase modulation driven by $m(t)$ is replaced with an integral that defines the phase response function. This integral ensures that $\varphi(t)$ remains continuous (as required for a CPM waveform), but discontinuities remain in its derivative $d\varphi(t)/dt \equiv \omega(t)$. As a result, large spikes occur in both the angular acceleration $\alpha(t) = d\omega(t)/dt$ and modulation torque $\tau_{mod}(t) = J\alpha(t)$. As described in [30], these can be removed by low-pass filtering the integral before phase modulation, resulting in

$$s(t) = A_c \cos\left(\omega_0 t + 2\pi \Delta f \times (n * h)(t)\right), \tag{4.4}$$

where $h(t)$ is the impulse response of the low-pass filter (LPF) and $n(t) \equiv \int_0^t m(t_1) dt_1$ is the filtered version of $m(t)$. This continuous-frequency FSK (CF-FSK) scheme ensures that both $\varphi(t)$ and $\omega(t)$ remain continuous, which in turn ensures that $\alpha(t)$ and τ_{mod} remain finite. The choice of $h(t)$ is relatively arbitrary. For example, a boxcar function $h(t) = rect(t/T_S)/T_S$ of width $T_S = 1/R_S$ can be used, which corresponds to a sinc function in the frequency domain. This particular filter is interesting because it keeps $\omega(t)$ piecewise linear, which in turn simplifies motor design by restricting $\tau_{mod}(t)$ to a finite set of values.

The value of $\alpha(t)$ decreases along with Δf. However, reducing Δf eventually destroys orthogonality between the symbols, which increases the bit error rate (BER). Assuming coherent or noncoherent detection, the minimum step size that ensures orthogonality is known to be $\Delta f = R_S/2$ or R_S, respectively, for CP-FSK. For simplicity, the same limit is assumed for CF-FSK.

Assuming constant Δf, increasing the number of symbols M also increases the torque required to transmit a message. Increasing the number of symbols increases the maximum and rms values of $m(t)$, thereby increasing the required rms motor acceleration (and therefore power) to transmit a symbol. However, each symbol represents $\log_2(M)$ bits and so increasing the number of levels increases the number of bits per symbol, thus decreasing the number of symbols required to transmit the same message. This trade-off results in an optimum value of M that minimizes the power required to transmit a fixed-length message.

The rms value of $\alpha(t)$ over the message duration, which sets the peak motor power rating, is represented by σ_α. Assuming the transmitted data is random, the rms angular frequency step is $\sigma_{\Delta\omega} = 2\pi \times \Delta f \times \sigma_D$, where σ_D is the rms value of the transmitted data values D. Since the symbol duration $\Delta t \equiv 1/R_s = 2/\Delta f$ for coherent detection and $\alpha(t) = d\omega(t)/dt = \Delta\omega/\Delta t$, σ_α is given by

$$\sigma_\alpha = \pi R_s^2 \sigma_D. \tag{4.5}$$

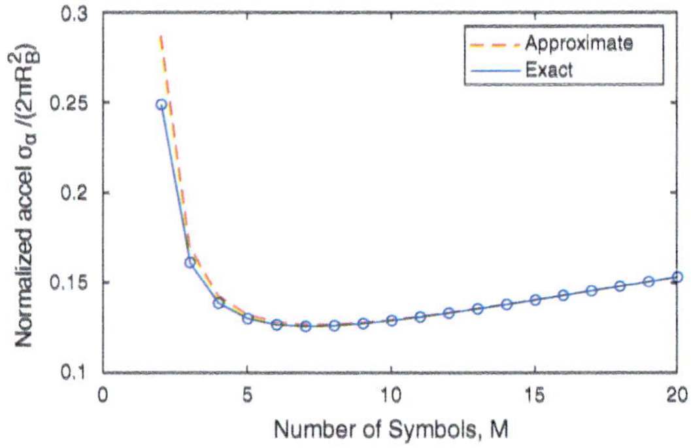

Fig. 4.9 Normalized value of rms angular acceleration versus the number of CF-FSK symbols

Each symbol represents $\log_2 M$ bits, so σ_α in terms of the bit rate R_B is given by

$$\sigma_\alpha = \frac{\pi R_B^2 \sigma_D}{(\log_2(M))^2}. \tag{4.6}$$

If we assume that each symbol is equally probable, σ_D can be approximated as $M/\sqrt{12}$ for $M \gg 1$. The normalized acceleration $\sigma_\alpha/(2\pi R_B^2)$ as a function of M predicted by an exact calculation of σ_D as well as the approximation is shown in Fig. 4.9. The optimal value is $M = 7$, which results in a $2\times$ reduction in σ_α compared to binary modulation ($M = 2$).

4.3.3 Simulation Results

Simulated CP-FSK and CF-FSK waveforms generated by the same random data stream are shown in Fig. 4.10. CF-FSK limits σ_α to $0.152 \times 2\pi R_B^2$, which is within 15% of the theoretical result. Figure 4.11 compares the power spectra of the two schemes estimated using Welch's method [31]. The figure shows that the transmitted bandwidth B_S is slightly smaller for CF-FSK than CP-FSK due to the additional low-pass filtering. The bandwidth for both schemes can be approximately derived by using Carson's rule for frequency modulation (FM) [32], which states that for $\Delta(f) \ll f_0$, $BW_S \approx (M\Delta f + 2f_m)$, where f_m is the bandwidth of the input signal. In this case, $f_m \approx R_S/2 = \Delta f$; we therefore expect that $BW_S \approx (M+2)\Delta f = 10\Delta f$ for $M = 8$, which agrees with the simulation results shown in Fig. 4.11.

4.3.4 Differential Coding

The proposed M-ary CPM scheme can be combined with differential coding to provide several benefits over conventional FSK. Each symbol α_i in such an M-ary scheme has a value in the range $\alpha_i \in [0, M-1]$ for M unique data symbol values. To increase the

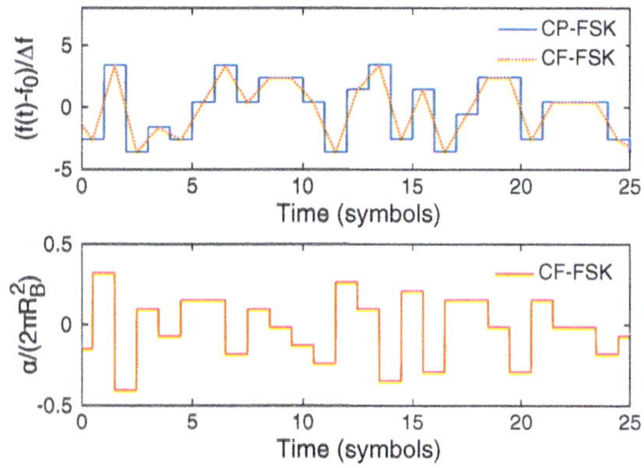

Fig. 4.10 Simulated instantaneous frequency offset (top) and modulation rate (bottom) for CP-FSK and CF-FSK, both with $M = 8$

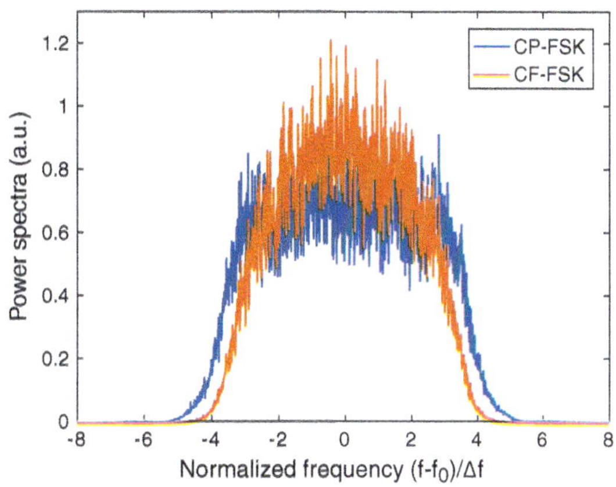

Fig. 4.11 Simulated power spectra of the transmitted waveforms for CP-FSK and CF-FSK, both with $M = 8$

number of bits transmitted per symbol, differential encoding is applied in a manner similar to differential phase-shift keying (DPSK). In a differential scheme, data is transmitted as the change from interval to interval rather than as the intervals themselves [24]. For example, DPSK is a binary scheme in which the symbols are encoded by the transformation

$$\alpha_i = |m(t_i) - m(t_{i-1})|. \tag{4.7}$$

A message transmitted with M-ary FSK requires a bandwidth $\Delta f(M-1)$ in the receiver as there are M fixed-frequency levels which must be observed by the receiver. However, consider the case of CPM with differential coding for the same value of M. Differential modulation schemes require the use of a special *differential detector*. For the case of DPSK, note the differential detector discards the sign of the difference between present and previous transmitted bits and records only the magnitude. For the proposed CP-FSK modulation scheme, using a differential detector sensitive to the magnitude as well as the sign of each transition, each symbol may take a value in the range $[-\frac{M-1}{2}, \frac{M-1}{2}]$. As a result, the required bandwidth for the transmission of M unique data symbols is only $\frac{1}{2}\Delta f(M-1)$. Thus, the required bandwidth for M unique symbols transmitted using M-ary differentially-coded CP-FSK (or CF-FSK) is half that of conventional FSK.

This type of differential coding is impractical to implement in conventional FSK as sign-sensitive differential detection of non-random data will inevitably require the transmission of long sequences of positive or negative values. In either case, no more data of the same sign may be transmitted once the magnitude $|m(t_{i-1})| = M-1$. The situation could be corrected by constructing a differential detector sensitive to transitions $m(t_i) - m(t_{i-})$ rather than $m(t_i) - m(t_{i-1})$ and adding additional frequency shifts within the transmit signal to code transitions which accumulate too great a net frequency shift over time. This frequency shift would be indistinguishable from a data transition in conventional CF-FSK modulation schemes, only it would occur some time within a symbol period. It is then critical that the differential detector is sensitive only to transitions at the data clock edge and not within the symbol period; otherwise, frequency discontinuities introduced to transmit successive differential-coded transitions of the same sign would be falsely detected as individual bits. In the proposed CF-FSK modulation scheme of Fig. 4.10, correction frequency shifts for diferential coding could be near instantaneous compared with the slow continuously time-varying frequency of each actual transition. A differential detector is readily adaptable to such a signal.

The differential detection scheme introduces some memory to the signal as each received bit depends on both the present and previous received values ($m(t_i)$ and $m(t_{i-1})$, respectively). As a consequence, one error in transmission produces two received bit errors, which doubles the BER [24]. The proposed CP-FSK scheme avoids increases in BER due to differential coding, as the differential detector uses the phase trajectory $\Delta\varphi(m(t))$ of the received signal rather than discrete present and previous values to determine the value of each bit. The phase memory introduced by the modulation scheme may then be used as additional information to increase the confidence of the detector in selecting the most probable current symbol.

4.3.5 Bit Error Rate (BER) Analysis

The BER analysis of CPM schemes is relatively complex. For the simplest case of an additive white Gaussian noise (AWGN) channel, the error probability of a maximum likelihood (ML) receiver for the ith transmitted symbol $s_i(t)$ is bounded by

$$P_{be}(i) \leq \sum_{k \neq i} Q\left(d_{i,k}^2 \sqrt{E_b/N_0}\right), \tag{4.8}$$

where $d_{i,k}$ is the Euclidean distance between $s_i(t)$ and the erroneous detected symbol $s_k(t)$; $E_b = A_c^2 T/(2\log_2(M))$ and N_0 are the energy per bit and noise power spectral density, respectively; and $Q(x) = \frac{1}{2}[1 - \mathrm{erf}(x/\sqrt{2})]$ where $\mathrm{erf}(x)$ is the error function [24].

The worst-case BER (i.e., upper bound) shown in Eq. (4.8) occurs when all $M - 1$ symbol errors are equally likely. Finding the upper bound is analytically difficult; instead, here the lower bound is considered. The latter can be derived by assuming that, though there are $M - 1$ possible ways to make an error per symbol, the receiver only makes the most likely errors, i.e., always selects the symbol that is "closest" to the transmitted one [25]. Experimental results confirm that this is a good approximation for the prototype ELF receiver. In this case, the minimum possible value of $d_{i,k}$ (denoted as d_{min}) sets the error performance of the receiver, and the total BER is approximately

$$P_{be} \approx Q\left(d_{min}\sqrt{E_b/N_0}\right). \tag{4.9}$$

The value of d_{min}^2 for the proposed CP-FSK modulation scheme is tabulated for some common values of M in [24, p. 66]. In particular, results are tabulated for $M = 4$ (quaternary) and $M = 8$ (octal) for various values of the modulation index h. The BER for the optimum case ($M = 7$) is expected to be bounded by these two cases; in the worst case, the expectation is $d_{min}^2 \approx 5.5$ for $h = 1$. The resulting BER curve versus E_b/N_0 is shown in Fig. 4.12.

In practice, an AWGN model is not accurate for atmospheric, underground, and undersea ELF and VLF channels [33, 34]. As described earlier in this chapter, atmospheric noise at such low frequencies exhibits impulse-like "sferics" induced by lightning strokes from storms, and "chorus" or "hiss" emissions from the magnetosphere, respectively; additional noise sources include power-line harmonics and ionospheric resonances. Many of these sources have time-varying and non-Gaussian probability distributions; they are also often correlated with fluctuations in the received signal amplitude (i.e., fading events). In addition to noise, underground and undersea propagation is frequency-dependent due to the dispersive medium. Thus, ELF channels exhibit both time- and frequency-dependent fading. The study of BER for CP-FSK or CF-FSK waveforms in such complex channels is a promising topic for future work.

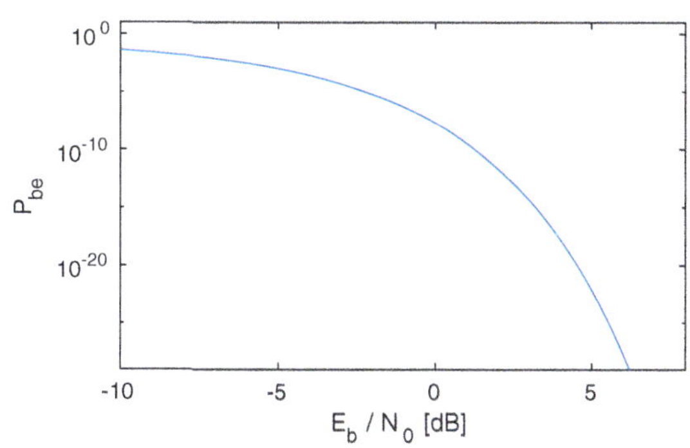

Fig. 4.12 Theoretical BER performance of a CP-FSK receiver in an AWGN channel

4.3.6 Data Encoding

Figure 4.9 shows that while $M = 7$ is the optimum number of CF-FSK symbols to minimize the torque due to modulation, the curve is nearly flat over the range [4, 15]. Moreover, the maximum frequency excursion for a given symbol rate is proportional to M. Regardless, we chose to use the optimum value $M = 7$. Assuming a differential detector, data with values in the range $[-3, 3]$ may thus be coded for transmission. This range is linearly mapped to [0, 6] for convenience. The DEFLATE algorithm operates on messages encoded using the ASCII character set and requires one eight-binary-bit byte per character. Transmission of this data using the seven available symbols would require three-septenary-bit bytes to represent all 256 characters in the ASCII set. However, proposed applications for ELF transmitters do not necessarily require the entire ASCII character set to be available. By implementing the coding of messages using a reduced character set (RCS) as described in Sect. 4.3.1, the effective data rate may be increased by reducing the number of symbols used to represent each character.

Consider a case where symbols are grouped into two-bit bytes. The numbers 00 to 55 in base-6 correspond to the numbers 0–9 followed by the Latin letters A–Z in base-36. This observation offers a convenient way to encode real data while leaving 14 additional values (when either MSB or LSB is 6 in a byte) for punctuation, data compression, and error-checking codes. In this character encoding scheme, each character uses two bytes, resulting in an effective data rate that is increased by a factor of $1.5\times$ compared to transmission using the ASCII character set.

The DEFLATE algorithm has not yet been implemented for such a reduced character set, so an alternative compression algorithm is proposed which operates on any character set. A combination of the Burrows-Wheeler transform (BWT) and run-length encoding (RLE) is used to achieve comparable compression ratios to DEFLATE in a direct comparison on ASCII data shown in Fig. 4.13. The Burrows-Wheeler transform applies block-sorting to rearrange the input message in a way that increases the frequency of runs of identical characters. Then, run-length-based compression is applied by replacing runs of identical characters in a string with one character followed by a counter to indicate how many times the bit is repeated. For example, a sequence such as "ABBBBC" may be replaced by "AB(3x)C". A single-byte counter code is used to represent the counter "(3x)" to minimize the number of bytes in the transmission-ready signal. Simulation shows this algorithm alone achieves a slightly lower compression ratio than DEFLATE. However, implementing a reduced character set after BWT+RLE compression results in an effective compression ratio above the performance of DEFLATE.

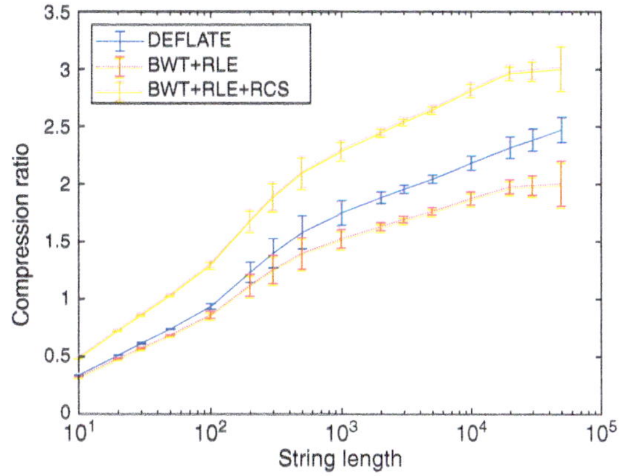

Fig. 4.13 Estimated compression ratio performance of DEFLATE, BWT+RLE, and BWT+RLE+RCS algorithms on typical English text (five well-known novels) as a function of string length

References

1. Persinger, M. A. (Ed.). (1974). *ELF and VLF electromagnetic field effects* (pp. 275–300). New York: Plenum Press.
2. da Silva, F. L. (2013). EEG and MEG: Relevance to neuroscience. *Neuron, 80*(5), 1112–1128.
3. Barr, R., Jones, D. L., & Rodger, C. J. (2000). ELF and VLF radio waves. *Journal of Atmospheric and Solar-Terrestrial Physics, 62*(17–18), 1689–1718.
4. Simões, F., Hamelin, M., Grard, R., Aplin, K. L., Béghin, C., Berthelier, J. J., et al. (2008). Electromagnetic wave propagation in the surface-ionosphere cavity of Venus. *Journal of Geophysical Research: Planets, 113*(E7).
5. Zhelavskaya, I. S., Spasojevic, M., Shprits, Y. Y., & Kurth, W. S. (2016). Automated determination of electron density from electric field measurements on the Van Allen Probes spacecraft. *Journal of Geophysical Research: Space Physics, 121*(5), 4611–4625.
6. Miller, A. B., Sears, M. E., Morgan, L. L., Davis, D. L., Hardell, L., Oremus, M., et al. (2019). Risks to health and well-being from radio-frequency radiation emitted by cell phones and other wireless devices. *Frontiers in Public Health, 223*.
7. Cohen, M. B., Inan, U. S., & Paschal, E. W. (2009). Sensitive broadband ELF/VLF radio reception with the AWESOME instrument. *IEEE Transactions on Geoscience and Remote Sensing, 48*(1), 3–17.
8. Singh, R., Veenadhari, B., Cohen, M. B., Pant, P., Singh, A. K., Maurya, A. K., et al. (2010). Initial results from AWESOME VLF receivers: Set up in low latitude Indian regions under IHY2007/UNBSSI program. *Current Science*, 398–405.
9. Crouchley, J. (1961). A Study of Whistling Atmospherics. *I. Occurrence. Australian Journal of Physics, 14*(1), 22–39.
10. Crouchley, J. (1964). A study of whistling atmospherics. V. Dispersion. *Australian Journal of Physics, 17*(1), 88–105.
11. Watt, A. D. (2013). *VLF radio engineering: International series of monographs in electromagnetic waves* (Vol. 14). Elsevier.
12. Cohen, M. B., Inan, U. S., & Paschal, E. W. (2009). Sensitive broadband ELF/VLF radio reception with the AWESOME instrument. *IEEE Transactions on Geoscience and Remote Sensing, 48*(1), 3–17.

13. Price, C. (2016). ELF electromagnetic waves from lightning: The Schumann resonances. *Atmosphere, 7*(9), 116.
14. König, H. L. (1974). Behavioural changes in human subjects associated with ELF electric fields. In *ELF and VLF electromagnetic field effects* (pp. 81-99). Boston, MA: Springer.
15. Mitsutake, G., Otsuka, K., Hayakawa, M., Sekiguchi, M., & Cornélissen, G., Halberg, F. (2005). Does Schumann resonance affect our blood pressure? *Biomedicine & Pharmacotherapy, 59*, S10–S14.
16. Cherry, N. (2002). Schumann Resonances, a plausible biophysical mechanism for the human health effects of Solar. *Natural Hazards, 26*(3), 279–331.
17. Ohayon, M. M., Stolc, V., Freund, F. T., Milesi, C., & Sullivan, S. S. (2019). The potential for impact of man-made super low and extremely low frequency electromagnetic fields on sleep. *Sleep Medicine Reviews, 47*, 28–38.
18. Rosen, L. A., Barber, I., & Lyle, D. B. (1998). A 0.5 G, 60 Hz magnetic field suppresses melatonin production in pinealocytes. *Bioelectromagnetics: Journal of the Bioelectromagnetics Society, The Society for Physical Regulation in Biology and Medicine, The European Bioelectromagnetics Association, 19*(2), 123–127.
19. Reiter, R. J., & Robinson, J. (1995). *Melatonin - your body's natural wonder drug*. Bantam Books.
20. Schmiedchen, K., Driessen, S., & Oftedal, G. (2019). Methodological limitations in experimental studies on symptom development in individuals with idiopathic environmental intolerance attributed to electromagnetic fields (IEI-EMF)-a systematic review. *Environmental Health, 18*(1), 1–24.
21. Eltiti, S., Wallace, D., Russo, R., & Fox, E. (2018). Symptom presentation in idiopathic environmental intolerance with attribution to electromagnetic fields: Evidence for a nocebo effect based on data re-analyzed from two previous provocation studies. *Frontiers in Psychology*, 1563.
22. Dieudonné, M. (2020). Electromagnetic hypersensitivity: A critical review of explanatory hypotheses. *Environmental Health, 19*(1), 1–12.
23. U.S. Energy Information Administration (EIA) [Online, last accessed 10.23.2020]. U.S. Energy Mapping System. https://www.eia.gov/state/maps.php.
24. Anderson, J. B., Aulin, T., & Sundberg, C. E. (2013). *Digital phase modulation*. Springer Science & Business Media.
25. Couch, L. W., Kulkarni, M., & Acharya, U. S. (2013). *Digital and analog communication systems* (8th ed.). Upper Saddle River: Pearson.
26. Salomon, D. (2004). *Data compression: The complete reference*. Springer Science & Business Media.
27. Ziv, J., & Lempel, A. (1977). A universal algorithm for sequential data compression. *IEEE Transactions on information theory, 23*(3), 337–343.
28. Tarek, M. T. B., Dharmasena, S., Madanayake, A., Choi, S., Glickstein, J., Liang, J., et al. (2018). Power-efficient data modulation for all-mechanical ULF/VLF transmitters. In *2018 IEEE 61st International Midwest Symposium on Circuits and Systems (MWSCAS)* (pp. 759–762). IEEE.
29. Goldsmith, A. (2005). *Wireless communications*. Cambridge University Press.
30. Tarek, M. T. B., Dharmasena, S., Madanayake, A., Choi, S., Glickstein, J., Liang, J., Mandal, S. (2018, August). Power-efficient data modulation for all-mechanical ULF/VLF transmitters. In *2018 IEEE 61st International Midwest Symposium on Circuits and Systems (MWSCAS)* (pp. 759–762). IEEE.
31. Welch, P. (1967). The use of fast Fourier transform for the estimation of power spectra: a method based on time averaging over short, modified periodograms. *IEEE Transactions on Audio and Electroacoustics, 15*(2), 70–73.
32. Carson, J. R. (1922). Notes on the theory of modulation. *Proceedings of the Institute of Radio Engineers, 10*(1), 57–64.

33. Bernstein, S. L., Burrows, M. L., Evans, J. E., Griffiths, A. S., McNeill, D. A., Niessen, C. W., Richer, I., White, D. P., & Willim, D. K. (1974). Long-range communications at extremely low frequencies. *Proceedings of the IEEE, 62*(3), 292–312.
34. Smith, A. J. (1995). VELOX: A new VLF/ELF receiver in Antarctica for the Global Geospace Science mission. *Journal of Atmospheric and Terrestrial Physics, 57*(5), 507–524.

5 Open-Loop Prototype

5.1 Prototype Experimental Setup

Diametrically polarized cylinder magnets are difficult to source from existing stock, whereas spheres are easy to find in the highest grade NdFeB material (known as N52). A sphere is the most efficient shape in terms of moment of inertia and volume per unit mass. These factors motivated an initial attempt to enclose a sphere in a metal shell which could then be attached to a rotating shaft.

This attempt proved inefficient; a first iteration in Fig. 5.1 shows the bulky housing required to mount a spherical magnet on a rotating shaft. The magnet is secured with compressible rubber o-rings on either side and could be adjusted to align the field as required, then torqued down from the other side using the six M3 hex bolts shown. The housing was later modified to be held together by a single compression-fit retaining clip which allows the mass of the shell to be significantly reduced and also reduces the bulkiness of the assembly.

In Figs. 5.2 and 5.3 the magnet has been affixed to two shafts mounted in blocks with rotating bearings, shown in free space as the design continues to evolve. The slip retaining ring allows for significant reduction in volume of metal in the shell while retaining the ability to secure the magnet with compressible o-rings.

The assembly in Fig. 5.3 shows the magnet and one o-ring and how they are held in place by the outer shell. This assembly was kept to produce a complete CAD design for the rotating assembly (shown in Fig. 5.4).

The challenges presented by this assembly are the continued issue of additional energy expenditure owed by the aluminum shell as well as the practical matter of manufacturing. The complexity of the aluminum shell makes it difficult to have manufactured. Reducing the complexity of machined parts aids prototyping speed and the total cost of the assembly. In the realized prototype, the sphere and shell are replaced by a ring magnet with a volume of 4.83×10^{-6} m^3 in grade N42 NdFeB. The lower volume and remanent field is a tradeoff

J. S. Glickstein and S. Mandal, *Electromechanical Transmitters for ELF/VLF Radio*,
Synthesis Lectures on Engineering, Science, and Technology,
https://doi.org/10.1007/978-3-031-15123-1_5

Fig. 5.1 Early attempts to contain a sphere within an aluminum shell. The sphere has a diameter of 25.4 mm and a volume of 8.58×10^{-6} m^3

Fig. 5.2 A spherical magnet

made to produce a power-efficient assembly where the cylinder magnet is mated directly to a shaft. In later work this cylinder is replaced with one of an optimum size.

An electronically commutated brushless DC motor (maxon EC6) was selected to drive the assembly as one of the limited options rated for speeds at or above 1 kHz (60 krpm), the initial target transmit frequency. This motor is shown in Fig. 5.5. At this early stage of development, the torque requirement to drive the assembly was not simulated and instead the motor was investigated separately from the rotating magnet, to evaluate control systems to drive the motor. In the completed prototype the nominal frequency was reduced and the motor replaced with a motor which delivers more torque but can only operate up to 166 Hz (10 krpm).

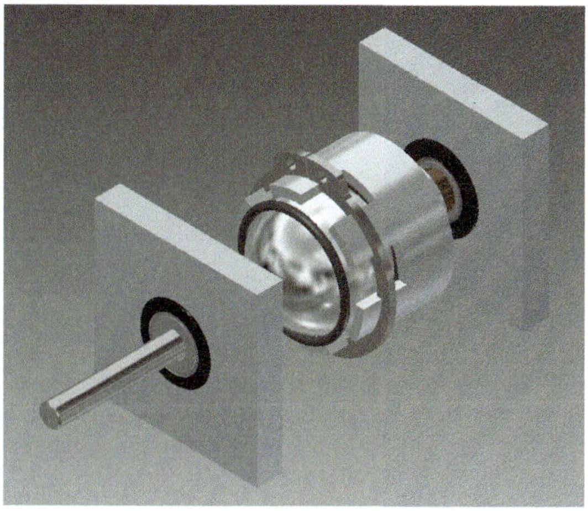

Fig. 5.3 The assembly of 5.2 with one half of the retaining shell removed, showing the magnet and one o-ring inside

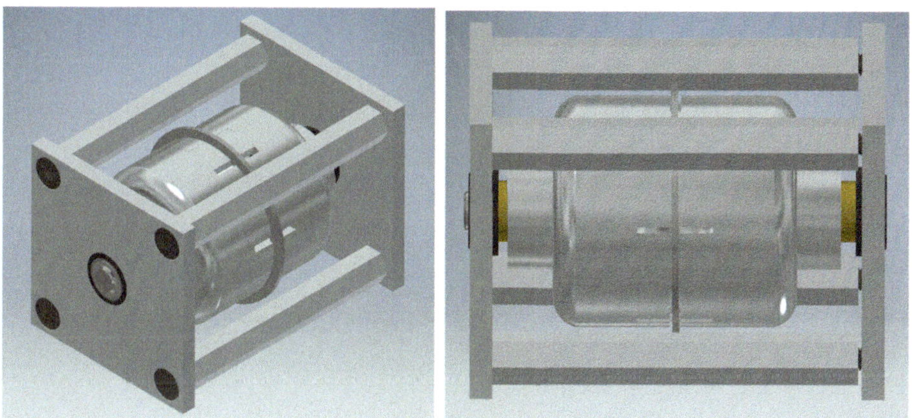

Fig. 5.4 The completed mounting system to house the rotating assembly is shown here. This proof of concept uses brass bushings

5.2 Prototype Experimental Setup: The Completed Assembly

An electro-mechanical transmitter designed to operate at ∼100 Hz was realized using a small brushless DC (BLDC) motor which drives a magnet made of grade N42 neodymium alloy (NdFeB) mounted to a shaft. The transmitter assembly is shown in Fig. 5.6. The motor is rated for a no-load speed of 10,000 rpm at 12 VDC, weighs 63 g, and is 34 × 33 mm ($L \times W$) in size with an output shaft 7.1 × 2 mm ($L \times D$). The magnet is a diametrically

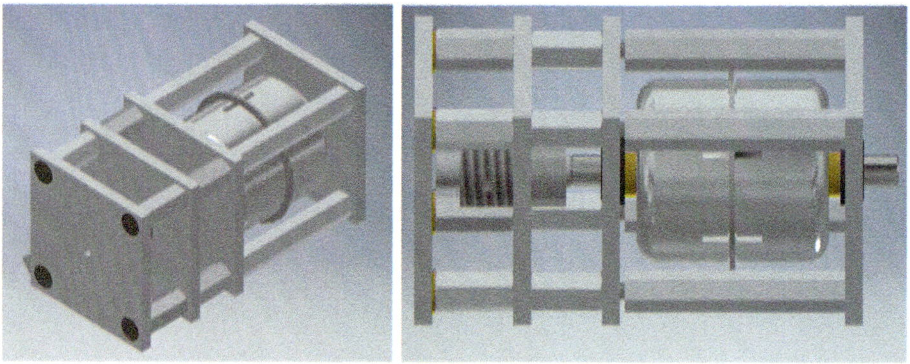

Fig. 5.5 The assembly is completed by addition of two brackets to secure the motor, which appears as a small grey cylinder in this rendering. The motor is held in place with a set screw in a hollow threaded plastic block

Fig. 5.6 A labeled photo of the prototype mechanical transmitter. Controller electronics are not shown in the photo

polarized ring with dimensions 19.05 mm outside diameter, 6.35 mm inside diameter, and 19.05 mm length. Given $B_r = 1.31$ T, the estimated magnetic moment is $m = 5.03$ Am^2. The complete assembly is $38 \times 38 \times 83$ mm ($H \times W \times D$) in size.

Design and control of the transmitter as well as an experimental setup for a receiver to test non-line-of-sight (NLOS) ELF communications are now discussed. Following tests of transmitter capabilities, experimental data transmission results are shown and data encoding/decoding methods are developed to meet environmental restrictions on the wireless channel. These restrictions include limited bandwidth, time-frequency resolution, and data rate.

5.2.1 Transmitter Design

The motor for the transmitter is driven by a power amplifier (PA) (AE Techron 7224) in voltage-output mode with open-loop control signals provided by a programmable voltage source (Keithley 2450 source-measure unit [SMU]) as shown in Fig. 5.7. Initially, the voltage-frequency characteristic of the motor was experimentally determined. This calibrated relationship was then used to calculate the SMU input waveform necessary for encoding user-specified data using the CP-FSK and CF-FSK modulation schemes described in the previous chapter. Finally, the SMU was controlled from a computer over a GPIB bus in order to run an experiment. With a suitably high Δf between input levels (such as 4 Hz used in the experiments which follow), errors from this open-loop control approach may conveniently be neglected. At low data rates (such as the 0.25–0.5 bits/s used experimentally), GPIB communication latency may be neglected as well.

5.2.2 Receiver Design

A block diagram of the receiver-side signal path is shown in Fig. 5.8. The receiver uses an inductive single-axis $\vec{B}$-field sensor (Magcheck-95 [MC95] from Magnetic Sciences), with a rated -3 dB magnitude bandwidth 25 Hz to 3 kHz. Interference from power lines poses a challenge to ELF communications; in the United States, such interference has a fundamental frequency 60 Hz. The initial goal was to transmit the signal between the 60 and 120 Hz interference lines. Thus, the initial transmit signals used in testing had $M = 15$ symbols and $\Delta f = 4$ Hz between levels for a data bandwidth of 88–116 Hz; the sensitivity of the MC95 sensor is effectively constant at 1 mV/mG over this range. Initially the receive-side signal path simply consisted of a low-noise amplifier (LNA) (SR560, Stanford Research Systems) with programmable gain connected to an oscilloscope. However, the LNA did not have a sharp enough cut-off to filter power-line interference without removing the signal itself, so

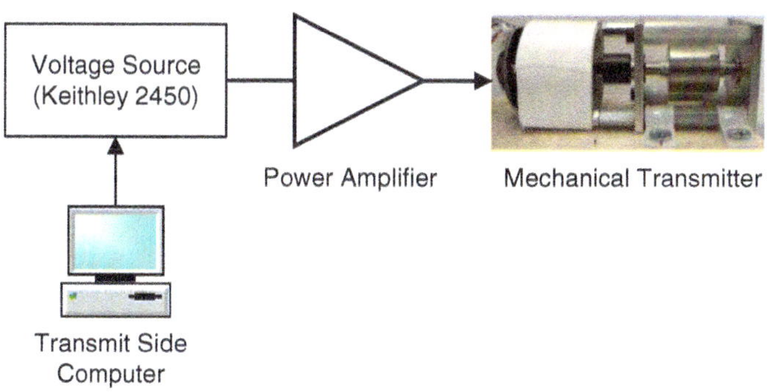

Fig. 5.7 Block diagram of the transmitter side of the experiment

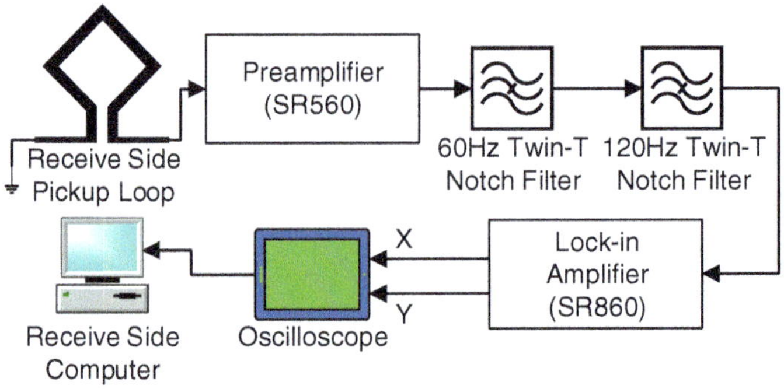

Fig. 5.8 Block diagram of the receiver side of the experiment

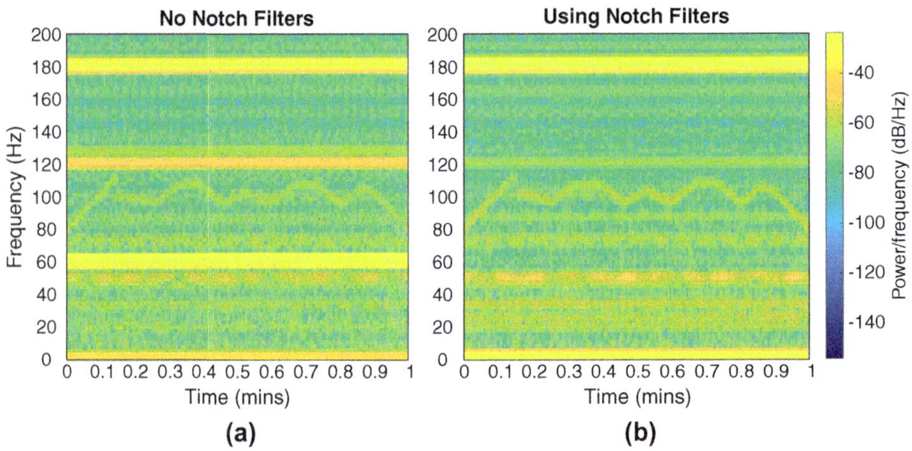

Fig. 5.9 Measured data at the input to the lock-in amplifier on the receive side at a location 4.67 m away from transmitter, shown **a** before, and **b** after the use of notch filters

two active twin-T notch filters (tuned to 60 and 120 Hz, respectively) were added to attenuate these sources. However, while transmitting across a distance greater than several meters, the LNA was found to quickly saturate before enough gain could be attained. To achieve longer transmission distances, a lock-in amplifier (SR860, Stanford Research Systems) with extremely high sensitivity (as low as 1 nV full-scale input range) was added after the notch filters [1]. The LNA was then used as a low-gain preamplifier to set the input-referred noise floor and buffer the input to the notch filters.

Figure 5.9 shows the spectrogram of a typical 10-bit CP-FSK sequence received with and without the use of notch filtering. The most invasive noise is the second harmonic 120 Hz, which is too close to the signal transmission bandwidth to be filtered out by the lock in amplifier. In Fig. 5.9a (without notch filtering), the LNA is set to a gain of 200, which was the highest achievable 60 Hz noise saturated the input to the lock-in amplifier.

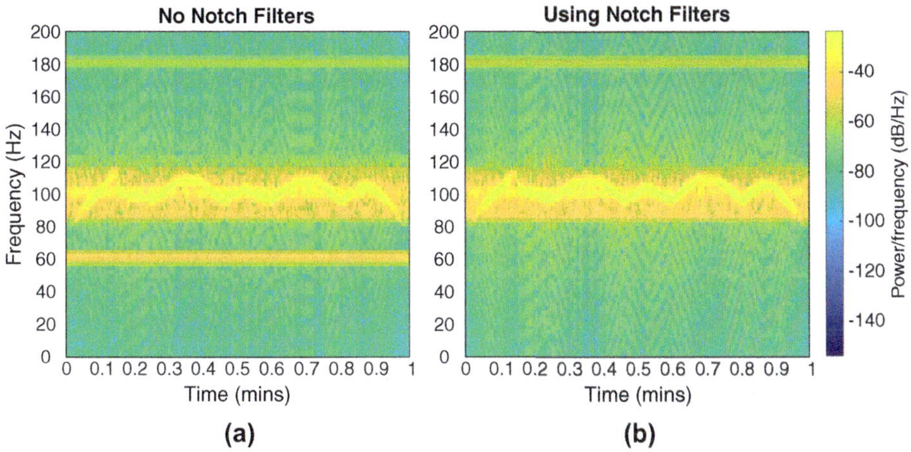

Fig. 5.10 Measured data at the output of lock-in amplifier on the receive side at a location 4.67 m away from transmitter, shown **a** before, and **b** after the use of notch filters

With the use of notch filters (see Fig. 5.9b), LNA gain could be increased to 500 before the input to the lock-in amplifier saturates, now by the third power line harmonic (180 Hz). The built-in filtering provided by the lock-in amplifier can reject this term, as shown in Fig. 5.10: The figure confirms that the third harmonic present at the lock-in output is significantly reduced (even for higher LNA gain) compared with Fig. 5.9. However, 180 Hz is still too close to the transmit signal bandwidth to be completely filtered out without also attenuating the transmitted signal. Thus, further improvements in receiver sensitivity are possible by using additional notch filters tuned to power line harmonics 180, 240 Hz, etc.

5.3 Experimental Results

5.3.1 Field Strength Versus Distance

In initial tests, the field strength versus distance d was measured at several points from 0.14 to 1.4 m away from the transmitter. The motor speed was kept constant 100 Hz (6 krpm) during these tests. The results, which are plotted in Fig. 5.11, confirm the expected field dropoff profile. In particular, the measured data is well fit by a function $\propto \frac{1}{d^3}$.

The receiver pickup coil has a sensitivity of 1 mV/mG (i.e., 10^4 V/T) over the transmission operating bandwidth, while the SR560 pre-amplifier has a noise floor of 4 nV/Hz$^{1/2}$. These values correspond to a measurement sensitivity of 400 fT/Hz$^{1/2}$, which is adequate for these tests. However, $\sim 10^3$ times better sensitivity is possible using state of the art $\vec{B}$-field receivers in this frequency range, such as the AWESOME instrument [2]. Given that $B \propto 1/d^3$, using such a sensitive receiver should immediately increase the useful range of the transmitter by

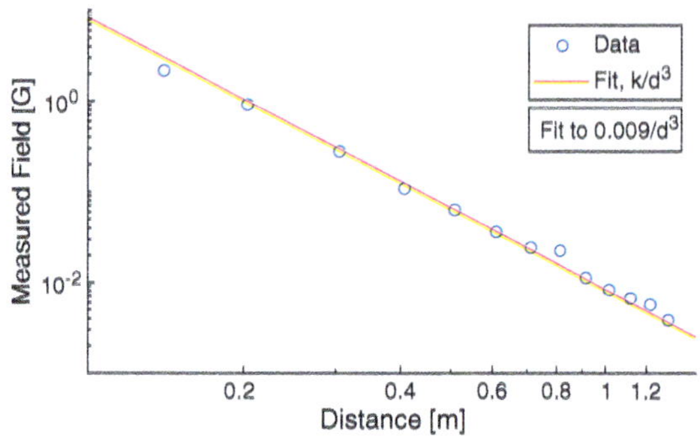

Fig. 5.11 Measured dependence of field strength versus distance 100 Hz

a factor of ∼10. However, the actual improvement in range is generally smaller due to the presence of atmospheric noise sources, as we shall see in later chapters.

5.3.2 Wireless NLOS Data Transmission

In order to transmit data, various CP-FSK and CF-FSK waveforms were generated in the manner described in Sect. 4.3.1. Using the experimentally determined frequency-voltage characteristic of the transmitter motor, transmit signals, such as those shown in Fig. 4.10, are converted to vectors of motor voltage corresponding to the desired transmit frequency. At low operating frequencies (typical center frequency 100 Hz with $\Delta f = 4$ Hz) and data rates (typical symbol period of 2–4 s), this rudimentary open-loop control system is sufficient to demonstrate successful transmission of data.

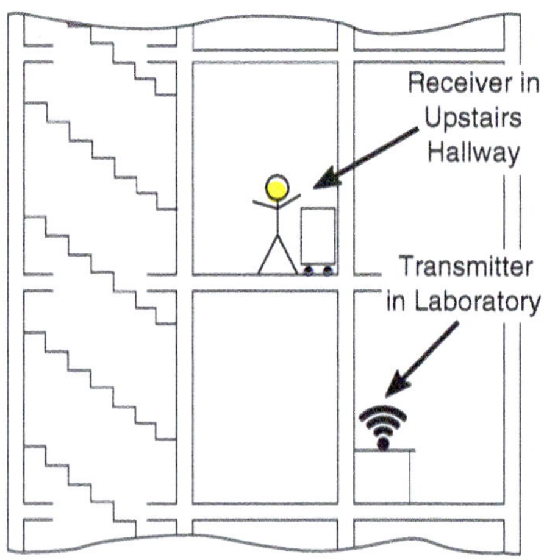

Fig. 5.12 Graphic depicting the location of the prototype mechanical ELF transmitter and inductive receiver in an office building. This setup was used for data transmission in the following sub-sections

Wireless NLOS communications using the mechanical transmitter was demonstrated by placing the receiver in a hallway one floor above the transmitter (see Fig. 5.12). The receiver was installed on a mobile cart in order to observe the ambient noise at various locations in the hallway. At the default location, the distance between transmitter and receiver in this setup was approximately 4.67 m with the two separated by a wall and floor/ceiling. During the tests, transmit data bandwidth was reduced to 96–108 Hz by using $M = 7$ symbols and $\Delta f = 4$ Hz. This narrow bandwidth permits more selective (larger Q factor) bandpass filtering in the lock-in amplifier to reject ambient noise and power line harmonics, as described in Sect. 5.2.2.

Figure 5.13 shows (i) example transmit data used to generate a 10-symbol CP-FSK message ($\approx$28 binary bits), and (ii) the reconstructed receive signal. The raw receive data values are plotted as a dashed line with the recovered bits shown as a solid line at the nearest valid bit value. The figure confirms successful data transmission.

The process of data recovery is now described. For convenience, a simple spectrogram-based decoder is implemented rather than the optimal MLSE-based decoder. Data is captured in the time domain and then post-processed using a short-time Fourier transform (STFT). Time windows of 1 s overlapping by 0.5 s result in an effective frequency resolution 1 Hz and a time resolution of 0.5 s, and generate spectrograms as shown in Fig. 5.14. Unlike the simulated sequences, the transmitted signal also includes ramp-up and ramp-down features which reduce wear on the motor and function as transmission "start" and "stop" indicators. Shown as a red trace in the figure is the peak of the PSD within each 0.5 s window, which is extracted to recover the transmitted signal.

The peak PSD shown in Figs. 5.14 and 5.15, with data points plotted on top of the smooth curve for clarity. The 10-symbol transmit sequence has a symbol period of 4 s, resulting in 8 sample points per bit for STFT windows of duration 0.5 s. These data points are split into bit periods as shown by light blue traces in the figure. A piecewise linear fit is used to extract the bit value in each region. Note that the data shown has a dynamic range of

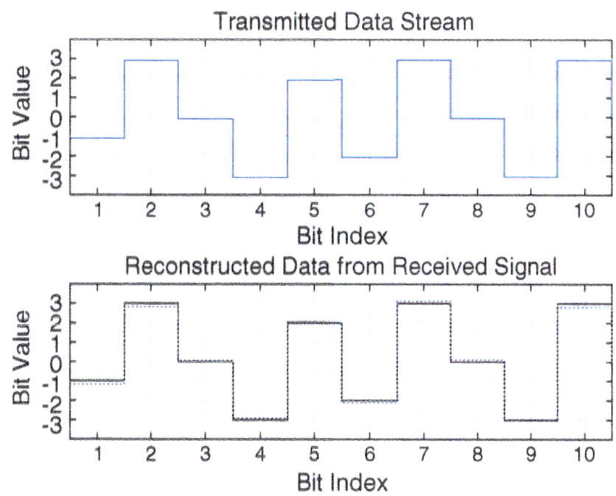

Fig. 5.13 Computer-generated transmit data bit stream compared with data received from the prototype transmitter

-3 to $+6$ Hz, while the transmitted waveform had a bandwidth -4 to $+8$ Hz. This error is due to open-loop motor control, and can be easily corrected by re-scaling this data before applying a linear fit since the received data still satisfies the condition $\Delta f \geq R_S = \frac{1}{T_S}$ for non-coherent detection, i.e., consists of an orthogonal set of symbols.

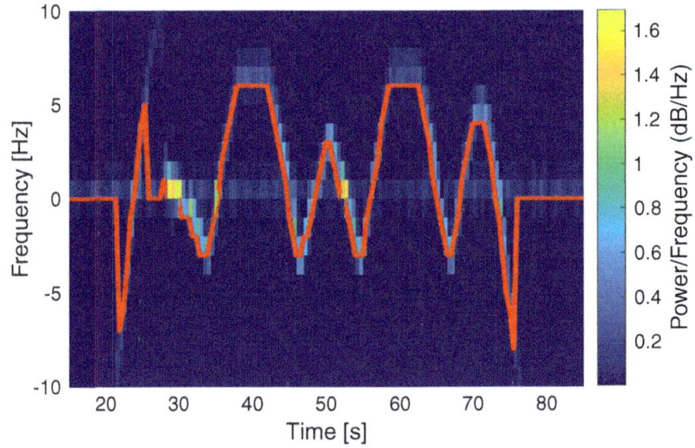

Fig. 5.14 Received power spectral density (PSD) of the signal versus time (multicolored, background) with peak intensity (red line, foreground) used to recover the data

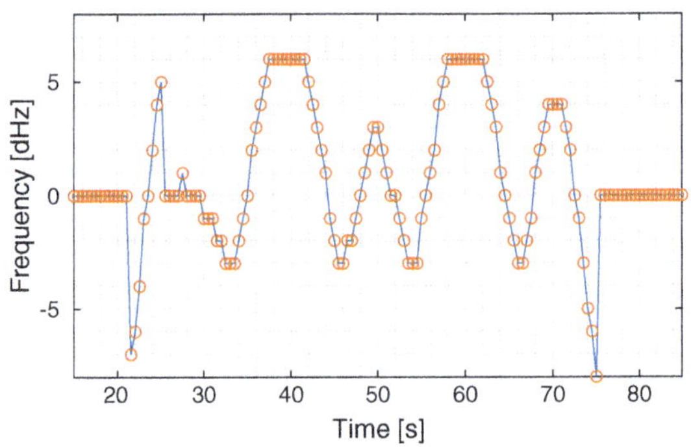

Fig. 5.15 Measured peak PSD plotted to demonstrate how each symbol is windowed to recover the transmitted data

5.3.3 Sample Data Encoding

Here is presented the message transmitted in the long-duration message transmission experiments, followed by the steps applied to produce the transmission signal $m(t)$. The transmitted message is the following 233-character excerpt from Shakespeare's *Hamlet*, Act II, Scene 2, with no new-line encoding.

```
WHAT A PIECE OF WORK IS MAN!
HOW NOBLE IN REASON,
```

```
HOW INFINITE IN FACULTY.
IN FORM AND MOVING,
HOW EXPRESS AND ADMIRABLE.
IN ACTION, HOW LIKE AN ANGEL.
IN APPREHENSION, HOW LIKE A GOD.
THE BEAUTY OF THE WORLD,
THE PARAGON OF ANIMALS.
```

The Burrows-Wheeler transform is applied to the data. The first step of this transform is to add beginning and end of sequence markers, shown here as '/' and '\' respectively. This increases the length of the output by two for a total of 235 characters to be transmitted.

```
/.ETNDESMNFNEWNNA,,,,!..E.
EWKWWSDWNYEEAN.,FFENGNNNDS
ELYD  R F RM M     PEHE OA
EAANNLOAKKHTLCHHLRBIRGHR O
OO N NN AWTTTE     PLLN
   VFSTM NRIIERBB  AURI D
IAIIIOIAOOOAAIIAAI ENG   G
SIIWWFMHHHHH  APXIA PPOOOS
ILNAEAI   CULCAOOOOOO\  ETT
```

Compression is applied using run length encoding (RLE). To visualize the effect, the run-counter character is written explicitly as (xN) where N is the number of times the previous character is repeated in the uncompressed sequence. Each counter is expressed as a single special character in the implemented reduced character set. The result is a sequence of 208 characters.

```
/.ETNDESMNFNEWN(x2)A,(x4)!.(x2)
E.EWKW(x2)SDWNYE(x2)AN.,F(x2)EN
GN(x3)DSELYD(x2)R F RM M(x5)PEH
E OAEA(x2)N(x2)LOAK(x2)HTLCH(x2)
LRBIRGHR O(x3) N N(x2) AWT(x3)
E(x5)PL(x2)N(x6)VFSTM NRI(x2)ER
B(x2) (x2)AURI D IAI(x3)OIAO(x3)
A(x2)I(x2)A(x2)I ENG(x3)GSI(x2)
W(x2)FMH(x5) (x2)APXIA P(x2)O(x3)
SILNAEAI(x3)CULCAO(x6)\(x2)ET(x2)
```

Finally each character is encoded to two bit bytes using septenary bits. The resulting stream is 208 bytes transmitted as 416 bits.

```
1606224535212244343523352252356114
3663460661220622523252614421523554
2261143506362361223524356221442233
```

```
5421606143602360433460346064412225
2260401422146135613340143261254533
2025613343153043242543604062603560
3561601452456222606441336135606551
2344453460354330612243156160611450
4330602160301430624030144062146130
6114613060223524606224443061526123
3425646061144153301460416140624430
3335142214306062205033201440652660
61224561
```

Using $T_s = 4$ s, this sequence takes 1664 s or 27 m 44 s. Implementing 8E0 (8 data bits, even parity bit, zero stop bits) would increase the size of transmissions by a factor of $\frac{9}{8}$, here resulting in 468 bits instead of 416 for a 12.5% increase in transmission time, up to 1872s (31 m 12 s).

5.3.4 Practical Message Transmission

Practical transmission of data through the link was demonstrated using the same 233-character excerpt from Shakespeare's *Hamlet*:

```
WHAT A PIECE OF WORK IS MAN!
HOW NOBLE IN REASON,
HOW INFINITE IN FACULTY.
IN FORM AND MOVING,
HOW EXPRESS AND ADMIRABLE.
IN ACTION, HOW LIKE AN ANGEL.
IN APPREHENSION, HOW LIKE A GOD.
THE BEAUTY OF THE WORLD,
THE PARAGON OF ANIMALS.
```

The full data encoding scheme shown in Fig. 5.16 was developed from the data modulation theory of Sect. 4.3.1 and preliminary experimental results. It was applied to the input data to produce a suitable transmission-ready signal. Compression reduced the sequence of 233 characters to 208 characters, 89% of the original size, as described in the previous subsection. For demonstration purposes, parity was not implemented and is left for future work.

The sequence of 208 characters was represented by 416 symbols using the established data encoding system. With a symbol period of 4 s, this data is transmitted in 1664 s (about 28 min). The spectrogram of data from the entire experiment in Fig. 5.17 shows an interesting consequence of open-loop voltage control. In a long transmission the transmitter bandwidth shifts, presumably as the motor heats up and coil impedance changes. Using a current source or switching to closed-loop control in future work is proposed to correct this problem.

Received data is processed using a demodulation scheme to recover the received message from the quadrature receive signal input to the block diagram of Fig. 5.18. The peak short time PSD of the entire received message is shown in Fig. 5.17. Windowing is applied to the peak PSD to identify the start and length of each symbol using the known symbol period and maximum rate of change in frequency. The piecewise linear fit produces a goodness of fit metric (R^2) which is used to determine whether the symbol window is reasonable.

Limited time resolution and errors present at transitions between symbols result in bit windows being less than the theoretical 9 samples which would represent a range of 4 s at 0.5 s intervals. In the absence of a clock signal, critical features such as peaks and discontinuities are used to identify the start position and length of each bit. These features are highlighted in Fig. 5.19, a sample 10 symbol sequence from the received message. This sequence also contains an error at index 378. The receive data decoding scheme of Fig. 5.18 shows two methods proposed to detect such an error. Using goodness of fit to detect an improperly windowed bit would not correct this data as this bit has a smooth profile and the linear fit computed goodness of fit is 0.9932. Implementing parity would allow for this error to be detected but not corrected. In either case, re-transmission of some data is required.

A short sequence of 10 symbols in the decoded receive signal is shown in Fig. 5.20. This sequence is chosen to correspond to the sequence shown in Fig. 5.19. In the transmission of 416 symbols, the error shown is the only erroneous symbol in the sequence. With retransmission of the short sequence including the erroneous symbol, successful transmission of the message is completed.

5.3.5 Bit Error Rate (BER) Analysis of the Prototype

Using the data collected in the previous section with the receiver placed 4.67 m away from the prototype transmitter, E_b/N_0 for the channel is measured to be 13.67 dB, which corresponds

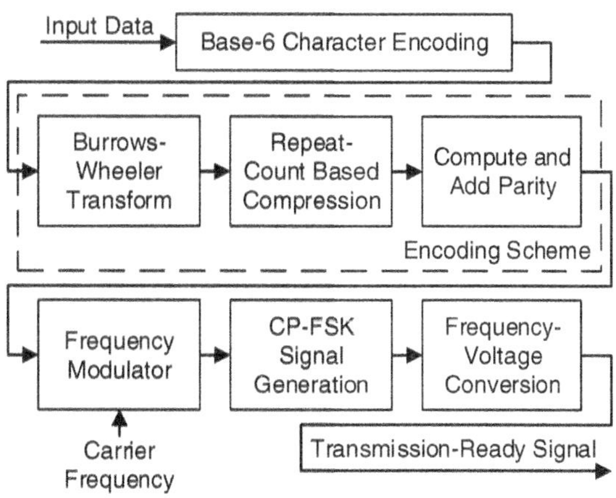

Fig. 5.16 Data encoding scheme used with prototype transmitter taking input data as a string of characters to produce a transmission-ready signal with compression and error-checking

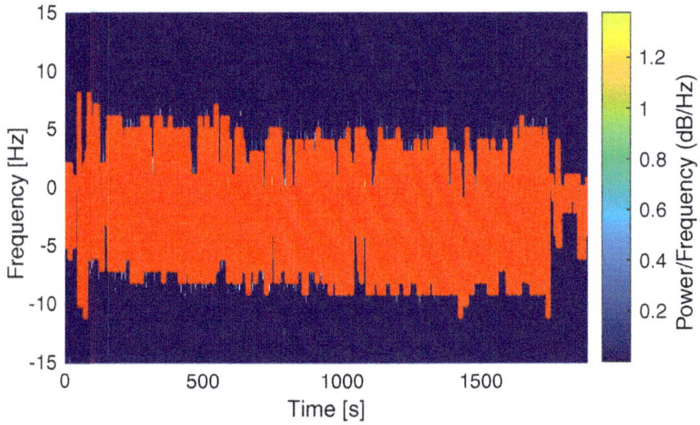

Fig. 5.17 Spectrogram of the received signal showing slow drift in transmission frequency over time

Fig. 5.18 Data decoding scheme used to recover the message from received data

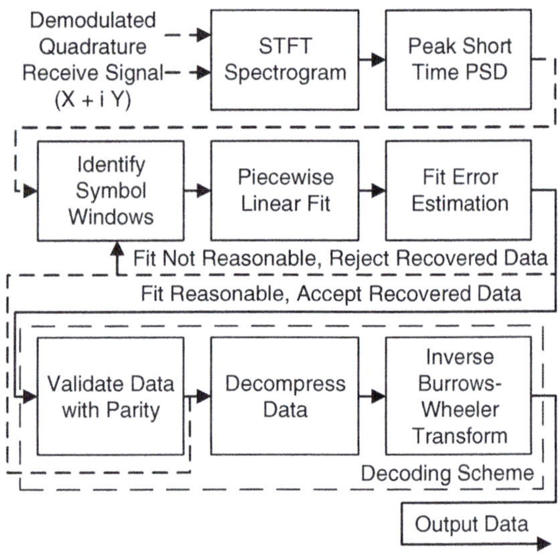

to a negligible BER of $P_{be} = 2.5 \times 10^{-22}$. In practice, a single symbol error was observed out of 416 symbols in the received message, which is promising but insufficient for estimating BER.

As magnetic field strength drops off as $B_{dip} \propto 1/r^3$ and $E_b = (A_s^2 T_s)/\log_2(M)$ in a non-conductive medium, the bit energy $E_b \propto 1/r^6$. This relationship is used to predict BER performance versus distance within such media by using the prototype mechanical transmitter. The field strength observed over distance in air (see Fig. 5.11) is consistent with simulations for a magnetic dipole. The expected BER performance of the prototype is extrapolated using this trend.

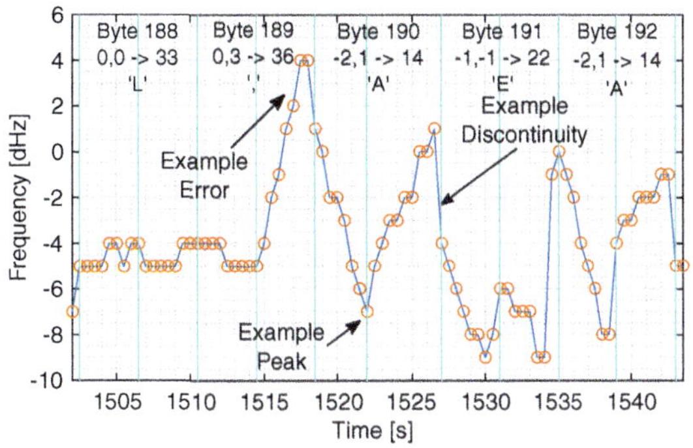

Fig. 5.19 Sample section of received CP-FSK signal with critical features and one error shown as an example

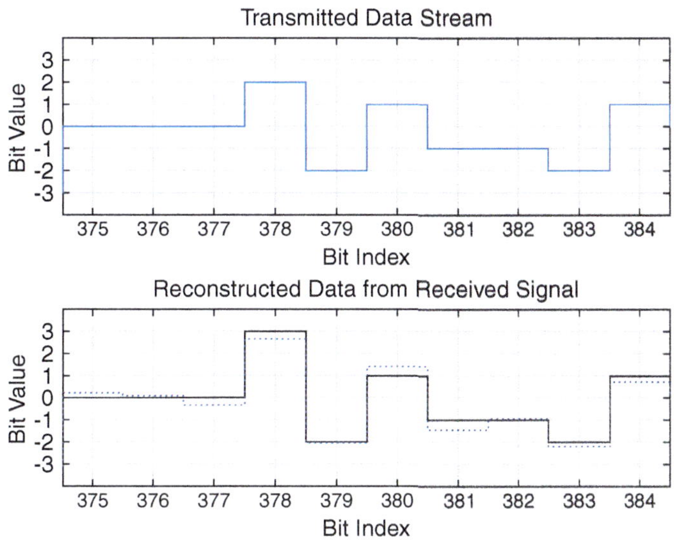

Fig. 5.20 Sample sequence of bits from message transmission experiment showing comparison between transmit and receive signals

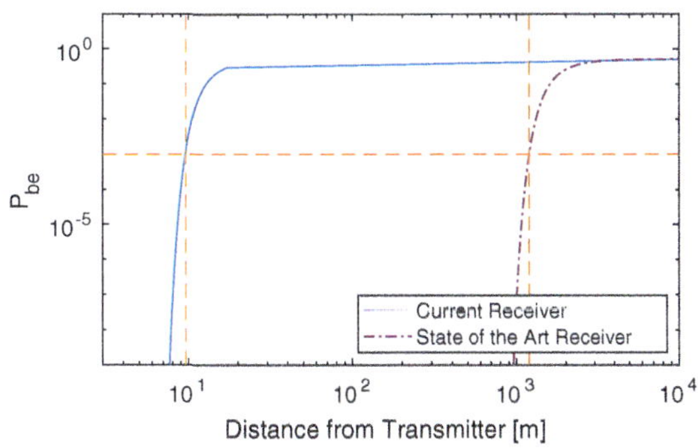

Fig. 5.21 Projected BER performance over distance based on experimentally determined E_b/N_0 in two cases: (i) using the current B-field sensor, and (ii) using the much larger AWESOME B-field receiver. The horizontal line denotes $P_{be} = 10^{-3}$

We define the theoretical maximum transmit distance as the distance where P_{be} reaches a value of 10^{-3}. This is generally considered to be a reasonable upper limit for message transmission without additional error correction. For the present receiver, this occurs at a distance of 9.6 m, as shown in Fig. 5.21. However, significantly longer-range communications are possible with a more sensitive B-field sensor. For example, consider the state-of-the-art ELF sensor used in the well-known AWESOME instrument [2]. This device uses a much larger detector (area = 1.69 m^2), resulting in a low input-referred noise of ~40 fT/$Hz^{1/2}$ at 100 Hz. In this case, the expected useful range of the transmitter increases to 1.19 km, as shown in Fig. 5.21. However, the actual improvement in range is likely to be smaller due to atmospheric noise, as noted previously.

5.3.6 Power and Energy Consumption with CP-FSK and CF-FSK

The energy consumption of CP-FSK and CF-FSK transmission was compared using the prototype transmitter. Three runs of 100 randomly generated symbols were transmitted with a bit period of 0.5 s and $M = 15$ symbols using both schemes. Each transmission was performed twice. Results are tabulated in Table 5.1. No significant difference between the schemes is observed in this data. This is because the non-optimized mechanical design of the prototype results in the load torque being much larger than the modulation torque (i.e., ensures that $\tau_L \gg \tau_{mod}$).

From the data in Table 5.1, average power consumption of the prototype mechanical transmitter is 11.0 W during operation. We compare this with the power required to generate an equivalent dipole field (with identical $1/r^3$ drop-off versus distance) using an electrical coil with the same dimensions as the prototype transmitter. In Fig. 5.22, we simulate the power required to generate an equivalent field with an electrical coil of varied fill factor. For fill factors of 0.4–0.9, the power required is about 75 W, or 6–7× the power used by the proposed prototype. These estimates are to be taken as lower bounds. The comparison does not account for the efficiency of the driver in each case, current crowding (proximity effect) or the thickness of the insulation in an electrical coil, or torque between elements in a distributed array of permanent dipoles.

Table 5.1 Comparison of energy consumption for 100-symbol transmit sequences using FSK, CP-FSK, and CF-FSK modulation schemes

Data set	Energy consumption (J)		
	FSK	CP-FSK	CF-FSK
1	1223	1171	1192
	1190	1191	1181
2	1183	1274	1265
	1177	1271	1264
3	1176	1132	1130
	1182	1130	1127

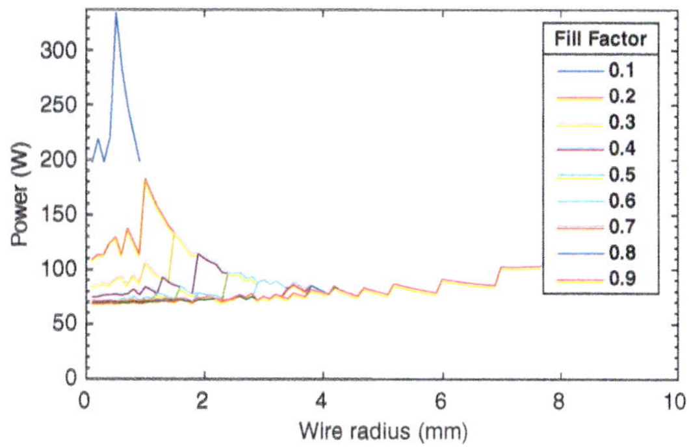

Fig. 5.22 Simulated power consumption (at 100 Hz) of an inductive coil with the same outer dimensions and magnetic moment as the mechanical ELF transmitter. Plots are shown for various values of fill factor ρ, where $1 - \rho$ is the ratio of inner to outer diameter for the coil

5.3.7 Effects of Non-random Data

Transmission of data by FSK modulation is affected by limited channel bandwidth. For example, in this case frequency deviations must remain smaller than ± 20 Hz to avoid 120 Hz interference. As a result, it is not possible to transmit long sequences of multiple bits with the same sign, as may occur in non-random data. To illustrate the problem, consider the case of the $M = 4$ level CP-FSK waveform used to transmit data with the prototype transmitter. The signal corresponding to five transmitted bits from the encoded excerpt of Shakespeare's Hamlet are shown in Fig. 5.23. The first bit is a $+3$, which may only be transmitted as a transition from the lowest FSK level to the highest level. At the highest level, a smooth transition could be used to transmit 0, -1, -2, or -3, but the next bit in the sequence is also $+3$. Therefore a discontinuity is required to limit transmit bandwidth, as shown. This behavior is compared to the ideal nature of the transition between bits 3 and 4 in the sample, which is a smooth transition. The transition from bit 3 to bit 4 would have lower τ_{mod} than the transition from bit 1 to 2.

Data transmission would be most power efficient if there were no frequency discontinuities such as those shown in Fig. 5.23. This is only feasible for random data. Non-random data will frequently require transmission of multiple repeated positive or negative values which results in discontinuities shown in the figure. This problem is visualized by plot-

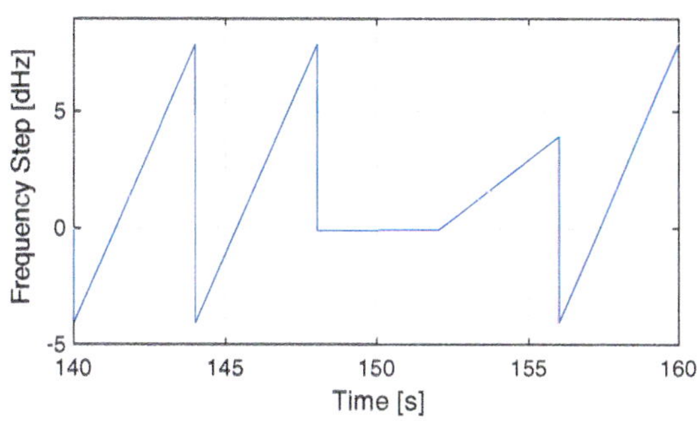

Fig. 5.23 Sequence of 5 non-random CP-FSK bits showing required frequency discontinuities for data transmission

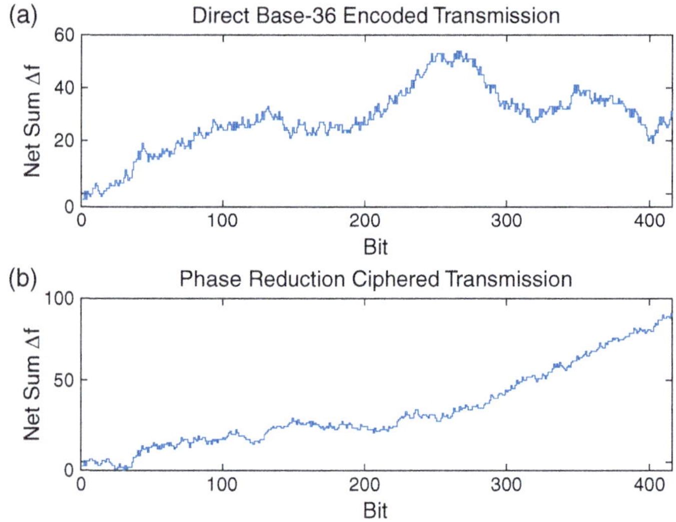

Fig. 5.24 Comparison of accumulated frequency shift for non-random data: **a** original data, and **b** with a phase reduction cipher

ting the accumulated frequency shift Δf across the number of transmitted bits in a sample 208-byte sequence of the previously described experiment. Figure 5.24a shows that the accumulated frequency shift can reach relatively large values. One proposed fix was to employ a substitution cipher before data encoding from base-36, assigning the most commonly used characters to average zero or low shift bytes. For example, we assign the most common letters in the English language, namely 'e', 't', and 'a', to transmit bytes (0, 0), (+1,−1), and (−1,+1) respectively. The result of applying this cipher is shown in Fig. 5.24b. Control and special character bits in the original encoding scheme were assigned to values beginning with +3, and were not changed by the cipher. As a result, the control bits cause a net positive accumulation of frequency shift over time which is not removed by the phase reduction cipher. However, the cipher is seen to be effective at removing some of the non-random shift caused by characters A–Z.

References

1. Scofield, J. H. (1994). Frequency-domain description of a lock-in amplifier. *American Journal of Physics, 62*(2), 129–133.
2. Cohen, M. B., Inan, U. S., & Paschal, E. W. (2009). Sensitive broadband ELF/VLF radio reception with the AWESOME instrument. *IEEE Transactions on Geoscience and Remote Sensing, 48*(1), 3–17.

6 Design of a Closed-Loop Transmitter

6.1 Main Design Goals

This chapter opens with a presentation of design constraints for the AMEBA2 in the mechanical and electrical domains. The electrical domain is abstracted and the system is treated as an entirely mechanical machine. We then reduce the mechanical model to a superposition of torques. Next, we develop a model for the optimization of thick-walled cylindrical ring geometries concerning power efficient modulation of angular velocity ω about the $\hat{z}$ axis. The model is used to identify the bounded region of operation for a mechanically based transmitter in terms of power consumption and field generation. By evaluating the optimization model for varying geometries, the impact of carrier frequency and modulation rate are put in context.

The design goals for AMEBA2 are outlined as follows:

1. Target carrier frequency f_c 100–1000 Hz
2. Ability to transmit an SMS-length text message in under 6 min ($D_{message}$)
3. Transmit distance ($d_{transmission}$) beyond 1 km
4. Error checking for message validation
5. Improved bit error rate (BER)
6. Improved power efficiency
7. Closed loop control of transmit frequency
8. Multi-cell redundancy
9. Completely integrated intelligent transmitter system control.

J. S. Glickstein and S. Mandal, *Electromechanical Transmitters for ELF/VLF Radio*,
Synthesis Lectures on Engineering, Science, and Technology,
https://doi.org/10.1007/978-3-031-15123-1_6

6.1.1 Bandwidth Selection

In evaluation and testing of the first prototype, one of the greatest issues in operation was the prevalence of 60 Hz noise and its higher harmonics. Notch filters at 60 and 120 Hz proved partially effective, boosting the effective range of the receiver from one entire meter to 4.67 entire meters. Beyond this receive distance, the 180 Hz harmonic saturated the first gain stage of the receiver preamplifier before any signal could be observed from the transmitter. While notch filters at higher harmonics are an option, the challenge of effective filtering can be reduced by moving to a higher frequency range as higher harmonics of 60 Hz noise are lower in intensity.

In further support of a higher bandwidth, note that receiver sensitivity improves with increased frequency since the voltage induced in a pickup coil is given by

$$V_a = j\omega N_a A_a B,$$

which is proportional to transmit frequency ω. Further analysis of receiver sensitivity will be discussed in Chap. 9.

6.2 A Model for the Mechanical System

6.2.1 The Fundamental Model

As the transmitted field is produced by mechanical rotation of a magnet, the conventional parameters of the realized communication system are directly related to the mechanical capabilities of the electromechanical transmitter. These parameters are accurately modeled from the perspective of the electric motor.

The communication system operates at a carrier frequency f_c. The peak torque requirement τ to produce acceleration in the rotating body, neglecting mechanical losses, is given by $\tau = \alpha \cdot J_{tot}$. Here α is the angular acceleration $\alpha = d\omega/dt = 2\pi df/dt$ of the motor shaft, where f refers to the rotation rate of the magnet in revolutions per second (which is equivalent to the frequency f of the electromagnetic wave produced by the mechanically rotated dipole magnet). J_{tot} is the total moment of inertia of the load, denoted by J instead of I for systems including an electric motor as I is defined as electric current. This relationship implies the maximum available carrier frequency f_c is fixed by not only the rated maximum speed of the motor, but also the available torque $\tau_{max,continuous}$ and the moment of inertia J_{tot} of the driven portion of the mechanical assembly.

For a symmetric set of M_i symbols where each symbol $m_i \in [-\frac{M-1}{2}, \frac{M-1}{2}]$, the worst-case (largest) acceleration of the motor would be required for a transition from the lowest level to the highest level. That is:

$$\alpha_{max} = \frac{\tau_{max}}{J_{tot}} = \frac{(M-1)\Delta f}{T_s} \tag{6.1}$$

This fundamental model represents a massively unconstrained optimization problem between the available motor torque, maximum rated motor speed, desired modulation rate, number of symbols M, and effective communication range (range $\propto E_s \propto J_{tot}$).

6.2.2 Constraining the System and Optimizing the Model

The fundamental model, despite appearing simple, reveals a great number of optimization parameters which seem to conflict with each other.

Decreasing the symbol period to satisfy the maximum message transmission period $D_{message}$ also increases the required torque τ_{req}. The increased torque inevitably requires a motor which cannot operate at the desired carrier frequency. Increasing the symbol period to achieve the opposite effect then fails to satisfy the maximum $D_{message}$.

There is a minimum required Δf, however the minimum value is not necessarily optimal given the relationship between Δf and BER on the receiver side. One could decrease M to decrease the required torque or allow for larger Δf without increasing τ_{req}, however $R_s \propto \frac{1}{M}$.

Increasing R_s to reduce τ_{req} also increases $D_{message}$. Decreasing symbol period results in the opposite effect, increasing the required torque to achieve the required message transmission period. However the possible torque on the motor is limited. Seeking a motor which may deliver the required torque results in the inability to achieve the required f_c.

Such analysis could be repeated *ad infinitum.* These analyses appear to run in circles, which suggests there exists more than one fundamental domain existing in direct relation to the other(s). Identifying these domains by sequential analysis of related model parameters is used to reduce the problem to a set of constraints based in a single principal determining factor resulting in a problem which may be solved. The solution is found through prioritizing model parameters and analyzing the factors which constrain each parameter to approach, then find, a single solution.

In M-ary FSK, the minimum frequency spacing to maintain orthogonality between carriers is $\Delta f_{min} = R_s/2 = 1/(2T_s)$ for coherent detection and $\Delta f_{min} = R_s = 1/T_s$ for noncoherent detection. Coherent detection requires a zero phase difference between symbols, noncoherent detection may be implemented for zero or nonzero phase difference between symbols. In the CP-FSK and CF-FSK modulation schemes developed for this transmitter, the phase difference between symbols is random and so only noncoherent detection may be used. Therefore the largest angular acceleration in Hz/s is given by:

$$\alpha_{max} = \frac{M-1}{T_s^2}, \tag{6.2}$$

and the corresponding torque required to produce this acceleration for the rotating body is given by:

$$\tau_{max} = \frac{M-1}{T_s^2} \sum J_n, \tag{6.3}$$

where J_n are the moments of inertia for each part of the rotating assembly. An in depth discussion of each component in J_n will be described later.

The optimum value of M is $M = 7$ for power efficient modulation has been established in Sect. 4.3.2.2, the same value used in the prototype. Therefore the maximum modulation torque required is given by:

$$\tau_{max} = \frac{6}{T_s^2} \sum J_n, \tag{6.4}$$

rewriting in terms of $R_{s,max}$ gives:

$$R_{s,max} = \sqrt{\frac{\tau_{max}}{6 \sum J_n}} \tag{6.5}$$

This expression for the partially constrained fundamental model elucidates the relationship between the communication system domain (communication domain) and the electromechanical transmitter domain (mechanical domain). The accidental philosopher remarks that this expression indicates these two domains are equivalent and must exist in harmony. All parameters of the communication domain boil down to an optimization of $\tau_{magnet\ assembly}$, the torque required to drive the magnet to produce the desired modulation. In the mechanical domain, the motor is limited by $\tau_{motor,max}$. In the electromechanical system, it is required that $\tau_{magnet\ assembly} < \tau_{motor,max}$ such that the motor is able to drive the magnet assembly.

6.2.3 Constraining the Communication Domain

An SMS message contains 160 characters. The encoding scheme described in Chap. 5 uses two symbols per character for 7-ary CP-FSK modulation without parity or flow control (8N0). Adding a parity bit (8E0), the required number of symbols is $N_s = 360$. To transmit this message in 6 min (360 s), a symbol rate of $R_s = 1$ symbol/s is required.

This design process intentionally ignores compression, which is not typically efficient for messages of such short length. In theory, using compression, the message could be transmitted in less than the allotted time. Along the same line of reasoning, the message data rate is expected to improve for longer transmissions.

Effective data transmission using the prototype is limited by two factors: frequency instability in the transmitter and low sensitivity of detection in the receiver. Both factors may be mitigated by increasing Δf, so a safety factor is wise in design of the transmitter to ensure Δf may be increased beyond the minimum value. Interestingly, a safety factor is also advisable for any mechanical system, and with the direct relationship between the communication and mechanical domains these safety factors are one and the same. Removing the restriction of $\Delta f = R_s$, the fundamental model is now:

$$R_{s,max} = \frac{\tau_{max}}{6\Delta f \sum J_n} \tag{6.6}$$

The solution reveals the principal limitations on maximum data modulation rate are the same as for the maximum carrier frequency, available motor torque and the inertial load of the driven portion of the mechanical assembly.

At a symbol rate of $R_s = 1$ symbol/s, the minimum Δf to maintain orthogonality is 1 Hz/s. As $\tau_{max} \propto \Delta f$, applying a safety factor of 2 in the mechanical domain also applies a safety factor of 2 in the communication system domain. Thus, a value of $\Delta f = 2$ Hz/s is selected. Substituting $R_s = 1$ symbol/s and $\Delta f = 2$ Hz/s in the partially constrained fundamental model yields:

$$\frac{\tau_{max}}{\sum J_n} = 12 \tag{6.7}$$

The fundamental model is now constrained in the communication domain and the result is an expression which requires constraint only in the mechanical domain.

6.2.3.1 A Brief Aside on Field Generation as an Optimization Parameter

Equation 6.7 is now fully constrained in the communication domain, however this result was only achieved by silently neglecting the desired minimum effective $d_{transmission}$ and BER. Minimum effective $d_{transmission}$ is limited by maximum available field amplitude $B_{dip,max}$ [1]. The maximum available field amplitude at the detector for a three-axis sensitive receiver is given in [1] as:

$$B_{dip,max} = \frac{2B_r V}{4\pi r^3} \tag{6.8}$$

For a single-axis detector, $B_{dip,max}$ is given by:

$$B_{dip,max} = \frac{(3\sqrt{3}+1)B_r V}{8\pi r^3} \tag{6.9}$$

Here $B_{dip,max} \propto V$, however $\sum J_n \propto V$ so constraining the model to achieve minimum $d_{transmission}$ would over-constrain the model by setting a torque requirement which cannot be satisfied, as will soon be explained.

The same challenge exists when considering BER, where $P_{be}(i) \propto E_b$ and $E_b = A^2 T_b$ (see Sect. 4.3.5). In the communication domain the symbol rate and therefore symbol and bit periods are fixed by the required maximum message transmission interval and cannot be increased. The only way to satisfy a minimum total generated field requirement in the developed model would be to increase the size of the magnet, which (given the model is fully constrained in the communication domain) would once again require an increase in the torque requirement in the mechanical domain. This is not fundamentally possible, as will soon be explained.

To understand these limitations it is necessary to understand the restrictions in capability of an electric motor.

6.2.4 Constraining the Mechanical Domain

The circuit model for a single coil DC electric motor is in essence an inductor with series winding resistance and a voltage source representing the memory caused by the inertia of the armature (rotor).

Figure 6.1 shows the model of an electric motor driven by a voltage source V. The three key components of the electric motor are the winding resistance R, coil inductance L, and back emf (or "motional emf") E.

The total torque developed by the armature acting on the rotor of an electric motor is given by $\tau = K_T \Phi I$, where K_T is a constant (set by the internal design of the motor), Φ is the total magnetic flux produced by the field from the current I in the armature wires, and I is the armature current [2, p. 77].

Back emf comes about from the rotation of the rotor, which induces a voltage (referred to as e.m.f. or E to distinguish this value from the applied voltage V of the motor driver) $E = K_E \Phi n$. Here K_E is a constant based on the influential factors in the design of the motor (number of conductors, length of the rotor, etc.), Φ is again the total flux in the armature, and n is the speed of the rotor (written as n to distinguish from velocity v).

In the majority of motors the flux Φ is constant at its full value and within the bounded operating region below the maximum speed-torque point. The relation $\tau = K_T \Phi I$ can then be rewritten as $\tau = k_t I$, and similarly $E = K_E \Phi n$ can be rewritten as $E = k_e \omega$ [2]. The constants k_T $[Nm/A]$ and k_e $[V/\omega]$ are equal in SI units, and are replaced by a single constant k. Making the replacement, these equations can be further simplified to:

$$\begin{aligned} \tau &= kI \\ E &= k\omega \end{aligned} \tag{6.10}$$

This is convenient for motor driver design as τ is directly proportional to I.

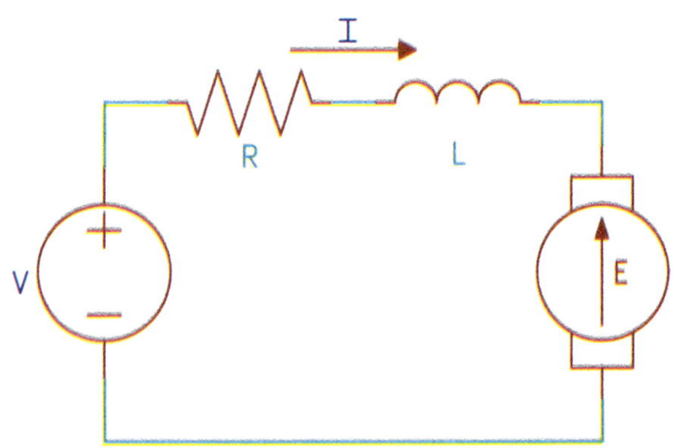

Fig. 6.1 Equivalent circuit of a single coil DC motor

Referring to Fig. 6.1, the voltage V must be greater than back EMF E in series with voltage drops across R and L, the armature voltage being described completely by the relation:

$$V = E + IR + L\frac{dI}{dt} \tag{6.11}$$

E has already been well described. IR is the voltage drop owed to series loss from winding resistance, and LdI/dt is the inductive voltage drop due to armature self-inductance. Armature self-inductance poses a challenge under transient conditions [2, p. 84], for example when the motor is at rest and must be driven at increasing speed to achieve operating speed ω_c (as a coupled system with a dipole magnet, angular velocity ω of the motor is equivalent to ω_c of the transmitted waveform). This may also cause difficulties in data modulation.

6.2.4.1 Transient (Dynamic) Behavior

When the electric motor is at rest, back emf is zero and applying a step function to reach the operating speed ω_c causes a massive current surge. This is deemed unacceptable in the literature [2] due to typical limitations in motor driver circuits, though a review of the available initial surge withstand current in high performance motors suggests that such state of the art motors typically allow for such a transient if the motor driver is able to supply such surge current.

The electric motor is a second order system with two time constants, a mechanical time constant and an electrical time constant [2, p. 98]. The literature references these time constants as the electromechanical time constant and armature time constant respectively, though motor manufacturers typically use the term mechanical time constant and leave it up to the engineer to determine the electrical time constant from winding specifications.

The mechanical time constant is defined by the current waveform $I(t)$ through the DC motor in response to a step function $\Delta V \cdot u(t) + V_i$, ignoring the armature inductance and assuming the motor is frictionless and unloaded. The resulting time constant τ_{mech} describes the response of the rotor speed and current, as shown in Fig. 6.2. The speed response is quite useful as a model for the motor's responsiveness, although the current response means very little given this is an idealized case study which does not account for the inductance of the armature windings.

The relationship between τ_{mech} and the motor/system parameters is $\tau = \frac{R_w J}{k^2}$ where J is the total rotary inertia of the motor plus load [2, p. 98], equivalent to the AMEBA2 system's $\sum J_n$, the sum of all load inertia terms in the mechanical system, and R_w is the winding resistance.

The electrical time constant is as one would expect for an inductive coil, $\tau_{elec} = L/R_w$. In a typical state of the art inductive DC electric motor, $\tau_{mech} \gg \tau_{elec}$. From the perspective of the motor, this allows analysis of the driven system to be considered a first order system relating to only τ_{mech}. The response τ_{elec} is important only in the design of the motor driver and determination of loop bandwidth.

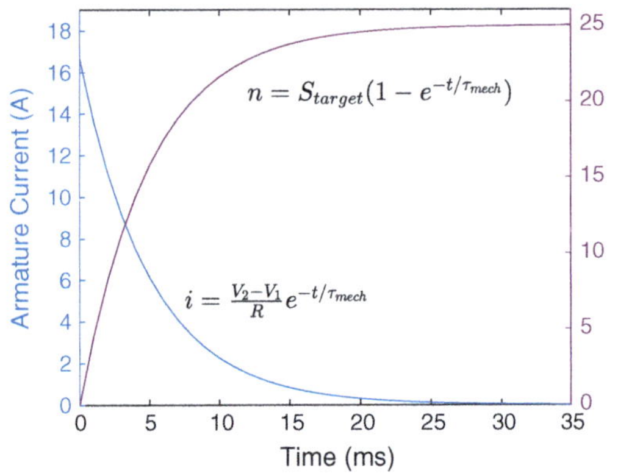

Fig. 6.2 Step response of an idealized DC electric motor. Consider n to be the rotor speed to distinguish this parameter from velocity v. This plot is generated with $V_0 = 0$ V, $V_1 = 5$ V, $R = 0.3\ \Omega$, $\tau_{mech} = 5$ ms, and motor speed constant $k = 5$ krpm/(min·V)

6.2.4.2 The Fundamental Constraint in the Mechanical Domain

The complete analysis of the electric motor now allows for the identification of the fundamental constraint of the entire system defined by the model in Eq. 6.1. The operating speed of an electric motor is set by the armature voltage, and the torque supplied is set by the armature current. In the steady state (constant V), neglecting armature impedance ($Z_a = R_w + \jmath\omega L = 0$ for analysis), torque and speed are related by

$$\tau \propto \left(\frac{V}{n}\right)^2 \tag{6.12}$$

where V is the voltage applied to the motor. This relationship indicates that the maximum torque decreases with operating speed. At low speeds, torque cannot be infinite, as $\tau = kI$ and I is limited by the winding resistance to $I_{max} = V/R_w$ (the thermal limit set by the motor design).

Figure 6.3 highlights two key regions of operation for an electric motor. Motor manufacturers specify the maximum rating of a motor as the values at the junction between the operating regions, namely the maximum capable operating speed under the conditions of full rated load and maximum voltage. In the constant torque region (at speeds below this point), the motor is capable of producing up to the maximum rated torque based on the design of the motor and the winding resistance. In the constant power region (at speeds greater than the rated speed), the motor can still operate but is not able to produce the rated torque. The maximum possible speed of the motor is in the maximum power region at the point where either the no-load armature torque is equal to the maximum torque the motor can deliver, or where the stator leakage inductance reduces the voltage on the armature to the point where it is no longer possible to motor to provide constant power operation, whichever comes first. In practice, this point is typically well beyond the point of thermal overload for the motor.

The constant torque region indicates the region where the motor is capable of producing maximum torque if required by the driven load; it does not mean that the motor will produce constant torque [2]. In an assembly the motor will supply the torque required to drive the

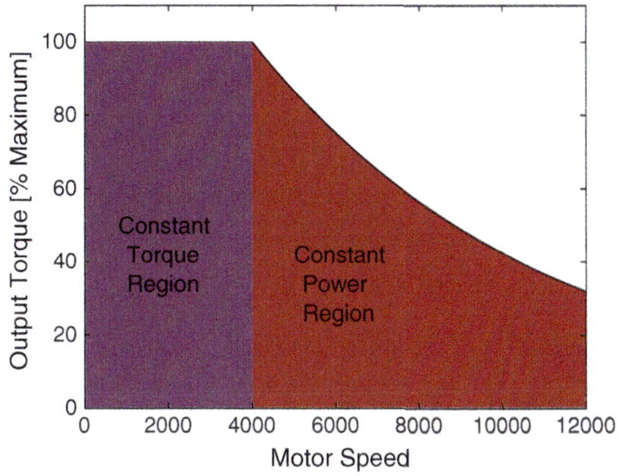

Fig. 6.3 A torque-speed curve for a typical electric motor showing the constant torque and constant power regions. In the constant torque region, torque is limited by winding resistance and flux saturation. In the constant power region, torque decreases proportional to the inverse of speed. This simulation assumes a motor rated for a maximum speed of 4 krpm at full load torque and a maximum operating speed of 12 krpm imposed by the motor designer. Output torque is normalized

connected load [2] up to the maximum capable torque. Speed may be varied by varying the voltage on the motor.

The constant power region is defined by the delivery of constant power rather than constant torque to the load. The idealized model of Eq. 6.12 risks confusion as (of course) it is well known that $P = \tau\omega$. A complete analysis of the electric motor requires added clarity in this area. By re-writing Eq. 6.11 for the armature voltage using the expressions in Eq. 6.10 for the developed torque and back EMF, the speed of the motor is given by

$$\omega = \frac{V}{k} - \frac{R_W}{k^2}\tau_L \tag{6.13}$$

Here the first term describes the no-load speed at arbitrary applied voltage V and the second term describes the reduction in speed due to a torque τ_L being delivered to the load. We can eliminate the motor constant in both terms of the equation and consider the equivalent torque in the first term as the torque required to maintain operation of the motor. In the second term we preserve the use of τ_L as a symbol for the torque delivered to the load. This gives

$$\omega = \frac{IV}{\tau_M} - \frac{I^2 R_W}{\tau_L}$$

It is easy to then identify electrical power as the numerator in both terms, which is no different from the equivalent mechanical representation for the system. Hence by abstraction,

$$\omega = \frac{P_M}{\tau_M} - \frac{P_L}{\tau_L} \tag{6.14}$$

In the case of the unloaded motor $P_M = \omega\tau_M$. Neglecting P_M and considering only the portion of the produced torque which is supplied to the load, $P_L = -\omega\tau_L$. From the perspective of the motor, the sign of P_M indicates P_M is delivered to the motor, and the sign of P_L indicates P_L is supplied by the motor.

6.2.4.3 The Constrained Mechanical Domain

Returning to the ponderance of Eq. 6.7, the required operating speed is fixed by the constrained communication domain. The system must be able to operate at the specified ω_c, and to modulate the carrier the system must operate within the range $\omega \in [\omega_c - \frac{BW}{2}, \omega_c + \frac{BW}{2}]$. Given the analysis of the electric motor (in particular the relationship between torque and speed as in Fig. 6.3), the required operating speed is fixed by the constrained communication domain and limits the possible torque for any selected motor. More succinctly, it's all about the torque.

6.2.5 The Gasoline Engine Analogy to the Electric Motor

An automotive engine produces torque curves as in Fig. 6.4, which increase up to a maximum torque $\tau_{engine,max}$ then fall as $\tau_{engine} \propto \omega_{engine}$. The parameters describing the functions $\tau_{engine}(\omega)$ and $P_{engine}(\omega)$ are well out of scope for this book and far too extensive to mention; suffice to say that much like an electric motor, the gasoline engine operates in two regions, one where the torque dominates (analogous to the constant torque region of an electric motor) and another where the power dominates (analogous to the constant power region of an electric motor).

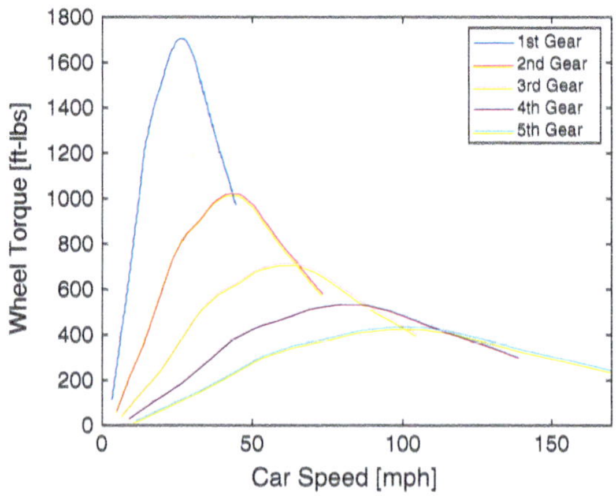

Fig. 6.4 The torque-speed curves produced by an engine producing 140 hp at 6000 rpm and 140 ft-lb at 4000 rpm, mated to a 5-speed gearbox with the standard single overdrive

In an automobile, the gasoline engine is mated to a gearbox allowing it to be capable of driving a load at a speed well beyond its own maximum operating point. This comes at a loss of torque delivered to the load. However, as $P_{engine} = \tau_{engine}\omega$ (the reader should now be fully convinced of this) and $n_{gearbox,output} \propto n_{gearbox,input}$, the gearbox allows the engine to continue delivering the same characteristic power curve. Carefully inspect the peaks of each curve in Fig. 6.4 and consider the power produced in each case. Further analogies to the gasoline engine are left to the imagination.

6.2.6 Summary of Constraints for Optimization

As described in Sect. 6.2.2, many variables are at play when defining the design parameters for AMEBA 2. Here is a concise summary of each design parameter and the variables which influence them:

1. Message Transmission Speed
    - Total message length
    - Baud rate R_s
2. Effective Transmit Distance
    - Number of unit cells ($V \propto N$)
    - Bit error rate
3. Channel Bandwidth
    - Maximum motor speed
    - Receiver sensitivity
    - Receiver SNR
    - Sources of noise within the selected band
4. Magnitude of Generated Field
    - Total volume of magnetic material
    - Number N of unit cells
    - Desired $d_{transmission}$ (effective transmit distance)

5. Size of Magnet Per Unit Cell

   - Total volume of magnetic material
   - Number of unit cells
   - Mechanical capabilities of an electric motor

6. Bit Error Rate

   - Magnitude of generated field
   - Desired effective transmit distance
   - Baud rate ($E_b \propto 1/R_s$)

7. Power Requirement

   - Power requirement of motor
   - Total volume of magnetic material
   - Air pump efficiency
   - Motor efficiency
   - Mechanical assembly efficiency.

6.3 Torque Required to Drive Assembly

The defining metric for the permissible volume of magnetic material in the system of rotating parts is the torque required to drive the system. The torque requirement at the input to the rotating assembly at any time t is taken to be the linear combination $\tau_{rot}(t) = \tau_{v(t)} + \tau_{\alpha(t)}$ of the static torque $\tau_{v(t)}$ and dynamic torque $\tau_{\alpha(t)}$. In this model, all loads and losses in the rotating system are modeled as a torque requirement. The value $\tau_{v(t)}$ is the torque required to maintain the current velocity, i.e., all sources of loss. The dynamic torque $\tau_{\alpha(t)}$ is the torque required to produce modulation, i.e., $\tau_{\alpha(t)}$ relates the acceleration of the rotating system to the torque required to produce such an acceleration. All loads coupled to the motor are now modeled as a torque requirement.

In Sect. 3.2, earlier work described the total torque requirement of the assembly as the two components τ_{mod} and τ_f for modulation and sources of loss respectively. Now a comprehensive analysis of both is undertaken to maximize the performance of the system. It will be shown that $\tau_{mod} = \tau_{mg} + \tau_{assy}$, the torque required by the magnet plus the torque required by the remainder of the components in the rotating assembly, and $\tau_f = \tau_{air}$ resulting from only air friction by using state of the art components and mechanical design. Power loss in non-mechanical components will be addressed in discussion of the electronics used in the control system.

6.3.1 Torque Due to Bearing Power Loss

The prototype AMEBA system used spherical roller bearings (colloquially, "ball bearing"), which cause several problems. Power loss in the bearings is high and this results in reduced torque available to rotate the magnet in the system, resulting in lowered field generation efficiency ($\vec{B}/P$ in units of T/W). The maximum speed rating of available ball bearings is well below the desired maximum carrier frequency 1000 Hz (60 krpm). Use of ball bearings requires a minimum pre-load force which increases with the maximum continuous speed rating of the bearing.

Total power dissipation in a ball bearing is given by

$$T_{loss} = W_T \frac{D_b}{60} + n_i M_{drag} \tag{6.15}$$

where W_T is the total energy dissipation of the bearing due to contact between the balls and both inner and outer raceways, D_b is the diameter of the balls, n_i is the rotational speed of the inner raceway, and M_{drag} is a variable which accounts for all other sources of energy loss in the bearing [3]. A specific solution requires application-specific parameters of the bearing to be simulated or experimentally determined in the completed design, including the internal load distribution in the assembly, which is not explored further [4]. Instead some typical values are explored (for completeness) as it will suffice to say $P_{bearing} > 0$ W.

An SKF 607 bearing is rated for a maximum rotational speed of 53 krpm ($\tilde{8}33$ Hz). This is the maximum rated speed which could be found through a survey of presently available ball bearings, well below the maximum rated speed in the AMEBA2 design specifications of Sect. 6.1. This bearing has inside diameter 7 mm and outside diameter 19 mm, making it theoretically suitable for use in a mechanical system at the scale of an AMEBA2 unit cell. However, roller bearings require a minimum load to produce the required traction in the bearing reaceways. If the unit cell uses two bearings to support the permanent magnet, the minimum required load (i.e., minimum mass of the supported assembly) is 2.63 kg at a rotational speed 100 Hz and 11.22 kg at 833 Hz. As will be soon shown in the final design of AMEBA2, this is unattainable for a system fully constrained by the model in Eq. 6.1.

Supposing the pre-load condition were attainable, the total power loss for two bearings under the condition of minimum allowable load is modeled using SKF's mechanical modeling software. The total power loss in the bearings estimated as 0.4 W 100 Hz (therefore $\tau_{bearings} = 4$ mNm) and 9.6 W 833 Hz (therefore $\tau_{bearings} = 11.5$ mNm). Both conditions reduce the mechanical efficiency of the system by an unreasonable margin, i.e., the remaining torque available to drive the rest of the rotating assembly including magnet itself is reduced to 75% or 28% in each given case respectively.

For power consumption and field generation efficiency (in addition to achieving up to the maximum target carrier frequency in the design specifications), air bearings can be used in place of ball bearings. Air bearings produce a thin cushion of pressurized air between the rotating shaft and the inside surface of the bearing, resulting in an effectively frictionless surface. The very small non-zero friction resulting from air friction is negligible in the realized assembly as justified in Sect. 6.4.1.

Manufacturers of air bearings are limited. One source of particular interest is the company OAV Air Bearing. Their OAV0250IB air bushing has an inside diameter of 6.35 mm (0.25″) and is rated for a maximum speed of 140 krpm. The manufacturer advised this bushing (in which the shaft makes direct contact with the porous media producing the air film) is suitable for the assembly and an air bearing (in which the shaft is supported by a raceway which makes internal connection to the air film) is not required.

Allowing the torque in the air bearings to be $\tau_{bearings} \approx 0$, the static torque τ_v is determined solely by air friction. Air friction $\tau_{air}(\omega) \propto A_{surface}$, so the general approach to next steps is to first model the general dimensions of the rotating assembly with an optimally sized magnet, then consider the resulting air friction and reduce the size of the magnet if required.

6.3.2 Modulation Torque

In the design of the real assembly, the maximum torque the rotating assembly may use is $\tau_{\alpha,max} = \alpha_{max} \sum J_n - \tau_{v,max}$ where α_{max} is 12 rev/s/s (as found in Eq. 6.7) and $\tau_{v,max}$ is the torque from air friction which for the sake of design analysis is neglected (at first). The components $J_n \in J_{tot}$ are:

- The rotating magnet
- A shaft coupling between the magnet and rotating shaft
- The shaft itself
- A shaft coupling between the rotating shaft and the motor (refer to Sect. 6.4.1)
- The armature of the motor.

The rotating magnet and the coupling between the magnet and rotating shaft are thick-walled cylindrical tubes with moment of inertia $J_z = \frac{1}{2} \cdot \rho V(r_i^2 + r_o^2)$ where ρV is the mass (in kilograms) calculated from density ρ and volume V to avoid confusion with the magnetic moment m of a dipole as used elsewhere. The shaft is a solid cylinder with moment of inertia $J_z = \frac{1}{2} \cdot \rho V r^2$.

6.3.2.1 Motor to Driven Shaft Coupling Specification

A shaft coupling is used between the motor and driven shaft in the assembly. A minimally sized shaft coupling is desired to minimize added moment of inertia. Additionally the shaft coupling must allow for some radial displacement such that the bearings in the motor and the

Fig. 6.5 Miniature mechanical couplings such as the bellows (left) and jaw (right) couplings are used in small rotating machinery

air bushings supporting the driven assembly are not loaded as a result of small yet nonzero misalignment of the two axes of rotation, and it must be balanced for high speed operation. Two options for the shaft coupling are shown in Fig. 6.5.

A bellows coupling is the ideal method to couple the motor to the driven portion of the assembly. These couplings exhibit high torsional rigidity for accurate transmission of torque and velocity between rotating assemblies while also remaining flexible enough to accomodate minor nonzero misalignment between shafts. The German company R+W Kupplengen manufactures customized miniature bellows couplings which can be made to meet all listed requirements, including mating ends drilled to US customary size on one side and SI units on the other. The specified coupling is balanced to a speed of 65 krpm to ensure balance is maintained for operation at (and slightly above) the maximum operating frequency of the assembly.

Jaw couplings are also well balanced and suitable for high speed applications. Compared with a bellows coupling, the jaw coupling allows greater misalignment as the two hubs of the coupling are mated through an elastomer insert called a spider. Axial misalignment can be corrected (zeroed) during setup of a device by repositioning the spider after jaw components on each mating shaft are tightened. Torsional stiffness and vibration damping are adjustable by interchanging spiders of various durometer. Increasing the durometer of the spider increases torsional stiffness and decreases vibration damping.

For the sake of analysis the moment of inertia contributed by this coupling is taken to be 1.1×10^{-7} kg·m^2.[1]

[1] Value taken from the datasheet for an R+W MK1 miniature bellows coupling with length 20 mm and outside diameter 15 mm.

6.3.2.2 Motor Inertia

Finally for the remainder of the models in this section the armature of the motor is taken to have moment of inertia 1.2×10^{-7} kg·m^2 (taken from Table 7.2).

6.3.3 Static Torque Due to Drag

AMEBA2 is designed to use air bushings in place of conventional rotary ball bearings. Air bushings are claimed to be utterly frictionless, an approximation which is accurate provided the air friction on the shaft and bushing is small relative to the air friction in the rest of the assembly. Given the intended size of the magnet assembly, this approximation is deemed acceptable. Friction inside the motor is accounted for by the specification of available torque and continuous operating point as the motor specification for maximum continuous torque is given at the output shaft and thermal dissipation in mechanical bearings is carefully set by the motor designer to allow continuous operation up to the maximum point of the continuous torque region. The only remaining source of loss in the assembly is due to air friction.

Despite the transmit frequency (100 Hz–1000 kHz) being remarkably low (so-called *extremely low frequency* in the communication domain), once again an interesting parallel is presented in the coupled mechanical domain, since the speed in the mechanical domain (6 krpm to 60 krpm) is remarkably high. For reference, the redline speed ($n_{max,continuous}$) for a 1978 Alfa Romeo Spider (see Fig. 6.6 for an aesthetically pleasing visual aid) is nominally 5.8 krpm, and this is a car with a mechancally-based engine designed to run extremely fast for what is considered typical. As a result, the effects of air friction in the AMEBA2 assembly are of particular concern.

Air friction is caused by drag on the rotating body. Much like the shear torque caused by the interaction between the rotating components of an engine and the surrounding oil film, air friction is actually a problem of wall shear stress between a moving solid (the assembly) and a viscous fluid (in this case, air) [5]. Suppose the cylindrical magnet were to rotate at

Fig. 6.6 The dashboard of a 1978 Alfa Romeo Spider

constant velocity about its ordinary axis of rotation. Wall shear stress τ_ω at the surface of the magnet is given by:

$$\tau_\omega = \mu \frac{du}{dy} = \mu \frac{du_\theta}{dr}, \tag{6.16}$$

where μ is the dynamic viscosity of air and the velocity profile u_θ is given in [6, p. 118] as:

$$u_\theta = \frac{r_0^2 \omega_0}{r} \tag{6.17}$$

Here the assumption is made that the outer cylinder does not rotate, that is, this analysis is limited by the assumption that the air around the cylinder is not pushed around by the rotation of the cylinder. This assumption is valid for an approximation where the rotating system is assumed to be at steady state given $\frac{d\omega}{dt} \ll \omega_0$ and for a sufficiently smooth outer surface such that the effect of the rotating body on the air around it is negligible. It is convenient to model air friction as a torque, consistent with the model of the mechanical assembly as a system which transmits torque from the motor to the rotating assembly. Torque $\tau = F \times r = F \cdot r$ along the axis of the rotating body. The force F acting on a given surface is related to shear stress as $F = \int \tau_\omega \cdot A$. Torque produced by shear stress from interaction with the air is then given by evaluating the integral separately for the two circular bases and the curved surface of the cylinder:

$$\tau_{air}(\omega) = 2 \iint \tau_\omega(r) r^2 \theta r + \iint \tau_\omega(R) R \theta L$$
$$\tau_{air}(\omega) = (4\pi \mu R^3 + 2\pi \mu R L)\omega \tag{6.18}$$

The model is limited by the assumption that air around the rotating body is stagnant. Sources suggest the boundary viscosity of air is increased by interaction with the rotating body [7]. This is attributed to momentum transfer from the rotating body to the air as a result of surface roughness. Acceleration of the system would be expected to increase this effect. Should air friction be experimentally observed as unreasonably high, it is expected that increased surface smoothness could decrease the influence of air friction.

6.3.4 Summary of Torque Requirements

All components contributing to the drive torque required by the mechanical assembly have now been identified. The lumped equivalent drive torque is therefore

$$\tau_m = \alpha(J_{mg}(R) + J_{assy}) + \tau_{air}(R, \omega) \tag{6.19}$$

where J_{assy} refers to $\sum J_n - J_{mg}(R)$, the total inertia of the rotating system less the inertia of the magnet itself which is left as a function of outside diameter (as is air friction).

6.4 The Optimum Magnet

6.4.1 Sizing the Magnet from Available Torque

It has already been established that field generation efficiency is maximized for long thin cylinders. Furthermore, it has now also been shown that both modulation torque and drag are minimized using such a geometry. The presented reasoning also applies (as an approximation) for thick-walled cylindrical magnets and the general trend is the same.

Figure 6.7 illustrates the relationship between field generation and torque required at maximum modulation rate for fixed length thick-walled cylindrical neodymium magnets as the outside diameter R is increased. The slope quickly tapers off as field generation increases far slower than the torque requirement.

In practice the optimization of the size of the magnet is (perhaps for the best) limited by manufacturing process capabilities for neodymium alloy (NdFeB) magnets. These magnets are typically produced by sintering which makes forming or drilling of long or thin holes difficult. Several manufacturers of custom dimensioned N52 grade neodymium magnets were consulted. The longest thick-walled cylindrical magnet which a manufacturer is willing to produce for AMEBA2 is 3″ and the smallest possible inside diameter of the hole is 0.25″.

Furthermore, the tolerance of such magnets (resulting from the sintering process followed by multiple layers of metal plating to protect the neodymium alloy from corrosion) from leading manufacturers is ±0.004″ for all dimensions. The magnet cannot be machined after manufacturing. Given the poor dimensional tolerance, in order to mount the magnet on the shaft, a coupling must be machined to fit between the precision steel shaft and the specific magnet used for each unit cell. Consulting a machinist, if the inside diameter of the coupling is a 0.25 in hole, the nominal outside diameter must be 0.5 in or greater.

These factors set the design limits before any further optimization could be considered. The maximum achievable length is 3 in, and the minimum hole size required to mount the magnet to a precision ground steel shaft is 0.5 in. These parameters are fixed in order to achieve high field generation efficiency, and the only parameter which remains to be selected is the outside diameter of the magnet.

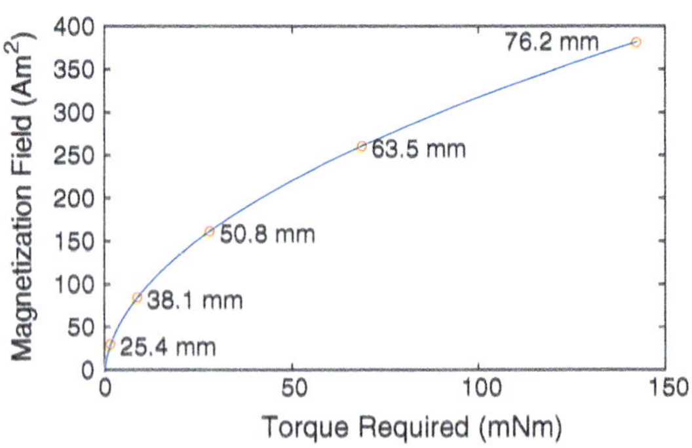

Fig. 6.7 Comparison of magnetization field produced and torque required to drive a cylindrical magnet of length 76.2 mm with a hole of inside diameter 12.7 mm. The outside diameter of the magnet is varied from 12.7 to 76.2 mm

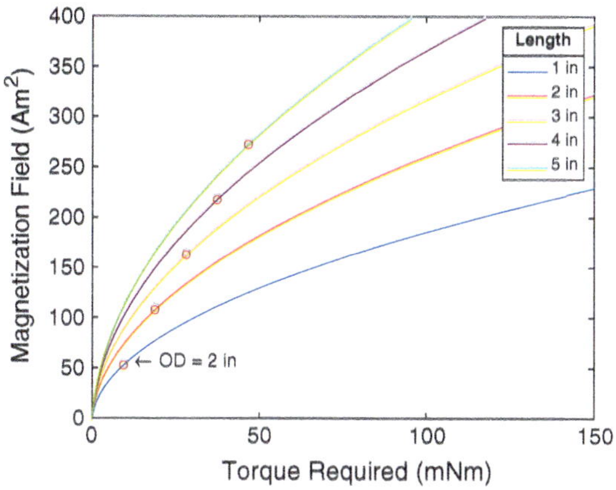

Fig. 6.8 Comparison of magnetization field produced and torque required to drive a cylindrical magnet of length 1–5 in with a hole of inside diameter 0.5 in. The outside diameter of the magnet is 0.5 in at the origin (i.e., an infinitely thin cylindrical tube) and increases along each curve. Fixed points on each curve mark points of constant outside diameter for reference. The scale of the plot is the same as in Fig. 6.7

The plot of Fig. 6.7 is reproduced in Fig. 6.8 with the length of the magnet now varied in increments as shown. Oddly enough, manufacturers of custom neodymium alloy magnets (all based in China) require dimensions specified in US customary units. Therefore Fig. 6.8 displays dimensions of length and outside diameter in inches.

According to the mathematical model, the most efficiently sized magnet in terms of power consumption has the smallest outside diameter achievable for any given length.[2] Despite this, it is inappropriate to select a minimally sized outside diameter as there are electrical power losses in the motor driver and in the motor itself, and mechanical power losses due to the inertia of other components in the mechanical assembly which do not contribute to field generation.

Accounting for all sources of loss in the system, a *mathematically optimum* sized magnet for the assembly is one which achieves the maximum possible field generation efficiency η_c. We consider the figure of merit for this condition to be defined as the maximum field generation referenced to maximum power consumption in the magnet $m/P_{in,max}$ (i.e., at maximum carrier frequency and peak modulation rate), which is convenient as this figure of merit is independent from data encoding and modulation schemes. A truly optimum unit

[2] If we were to construct a completely lossless motor (cryogenic cooling of the armature windings can eliminate electrical power loss) and float the magnet between air bushings without the support of a shaft, this would truly be our answer. This is perhaps best reserved for future work.

cell in terms of field generation efficiency would then use a magnet sized such that J_{mg} results in the condition of maximum power transfer:

$$w_c\alpha_{max}J_{mg} = P_{in}(1-\eta_{driver})(1-\eta_{motor}) + w_c\alpha_{max}(\sum J_n - J_{mg})$$

It is convenient to express this relation in the entirely mechanical domain, which furthermore eliminates the frequency dependence. The lower bound for field generation efficiency in the assembly occurs when

$$J_{mg} = \frac{1}{w_c \cdot \alpha_{max}} P_{in}(1-\eta_{driver})(1-\eta_{motor}) + \sum J_n - J_{mg} \tag{6.20}$$

This condition could easily be satisfied if the number of unit cells were limitless. In practice, a greater number of unit cells requires increased space for the total assembly, increased cost, and furthermore increases the complexity of the control system.

6.4.1.1 Practical Optimization

In practice the system must also not take up excessive space or be unreasonably expensive to manufacture, so instead an approximation is used. Given the length and inside diameter of the magnet are fixed at 3 in and 0.5 in respectively, the field generation per unit torque required are now based on the outside diameter of the magnet. Consider the yellow curve of Fig. 6.8 which models this trend. Increasing the outside diameter improves the net magnitude of the magnetization field at relatively low cost to the torque requirement for outside diameters less than 2 in. Beyond this point, there is a continually diminished increase in magnetization field per unit increase in required torque. Therefore given the tradeoffs between truly optimum field generation efficiency and optimum field generation efficiency *in a realistic, practical assembly*, the OD of the magnet should be less than 2 in.

Of course it remains to be seen exactly how large the outside diameter of the magnet may be given the available torque supplied by the motor. Before this may be completely elucidated, one more model is required, one which considers the effect of air friction in the rotating assembly, which will of course further increase the torque required as the size of the magnet (and therefore its surface area) is increased.

6.4.2 Real Size of the Magnet, A Boundary Value Problem

Using the developed models for all torques, the upper bound for the load on the motor is given by

$$\alpha_{max}(J_{mg}(R) + J_{assy}) + \tau_{air}(R, \omega_{max}) \leq \tau_{m,max} \tag{6.21}$$

For boundary analysis, the length of the shaft is set to 8 in to allow for a generous clearance between the motor and supports. The length and inside diameter of the magnet are chosen as

3 in and 0.5 in as the final parameters. Solving symbolically yields the maximum allowable outside diameter of the magnet, 1.7034 in (the outside diameter at the lower bound of η_c).

Analysis of specific moments of inertia informs $J_{mg,max} \gg J_{assy}$, which allows further simplification of Eq. 6.21. While this simplification is inadvisable as the specific solution provides maximum accuracy and allows for simulation of assemblies with arbitrary dimensions for each component, it does inform that the solutions remain accurate even if the length of the shaft changes in the final assembly.

Assuming the power consumed by the motor driver is dominated by the output stage MOSFETs and the worst case $R_{DS,ON}$ is 0.2 Ω, $P_{drive,wc} = 0.4(\tau_m/k)^2$ for any τ_m. Similarly, $P_{rotor,wc} = 0.208(\tau_m/k)^2$. The minimum size of the magnet for an assembly which is *optimum in practice* may be calculated as a symbolic solution to Eq. 6.20 under these conditions as

$$J_{mg}(R) = \frac{1}{\omega_{max} \cdot \alpha_{max}}(P_{drive,wc} + P_{rotor,wc}) + J_{assy} \tag{6.22}$$

where in the symbolic solution τ_{air} is modeled as its equivalent moment of inertia as a member of J_{assy}. Solving Eq. 6.22 symbolically yields the defined minimum allowable outside diameter of the magnet, 0.8186 in (the outside diameter at the upper bound of η_c). At this point the field generation efficiency is at its mathematical optimum (in reference to the figure of merit described in Sect. 6.4.1). Below this point, power loss in other components of the assembly P_{loss} are greater than the power requirement of the magnet and net field generation in the cell is reduced. There is no benefit to further reducing the modulation torque in this scenario hence the solution to Eq. 6.22 is the lower bound on the practically optimum outside diameter of the magnet.

Above this point $P_{mod,max} > P_{loss}$ and the magnitude of the field generation increases $\propto OD$, though field generation efficiency drops. This trend is shown by the gradient of Fig. 6.8, indicating field generation efficiency of a single cell as the size of the magnet is increased beyond the optimum value. In this case field generation efficiency is sacrificed to improve the overall efficiency of the assembly in terms of the practical parameters of cost, size, and complexity, the unit cell is one of many components which comprise the complete transmitter system.

Torque requirements in the assembly for various selections of OD are shown in Table 6.1 for ease of reading. The shaft length is assumed to be 6.5 in though its contribution has no noticeable effect on the results; values 4–8 in contribute <0.005 mNm.

Figure 6.9 shows η_c as a function of varied outside diameter for the proposed magnet. The red dashed line shows the upper bound on the outside diameter of the magnet OD_{max}, where torque required by the assembly under worst case conditions is equal to the maximum continuous torque available from the motor selected for analysis (16 mNm). Assuming $\tau_{motor,max}$ is fixed, in reality the green efficiency curve does not extend beyond this position,[3]

[3] There are methods to exceed $\tau_{motor,max}$ though curve would look a great deal different and this is beyond the scope of our conversation.

Table 6.1 Tabulated values of torque $\tau_{mg} + \tau_{assy}$ required to drive mechanical components of the assembly, τ_{air} at the bounds of the defined channel bandwidth, and τ_{req} the total torque the motor must supply to the assembly at peak modulation rate

Magnet OD (in)	Load torque (mNm)			
	$\tau_{mg} + \tau_{assy}$	τ_{air}, 100 Hz	τ_{air}, 1 kHz	τ_{req}
0.8186	0.8	0.18	0.59	1.3
1	1.7	0.22	0.72	2.4
1.2	3.6	0.26	0.86	4.5
1.4	6.7	0.30	1.00	7.7
1.6	11.5	0.34	1.10	12.6
1.7034	14.8	0.37	1.20	16.0

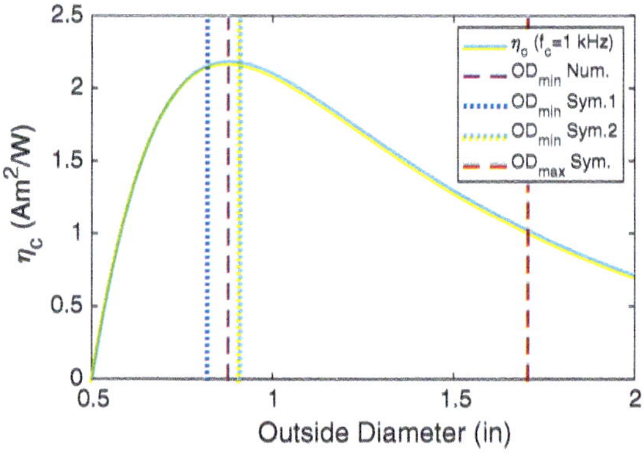

Fig. 6.9 Field generation efficiency at peak modulation rate and maximum carrier frequency 1 kHz for the proposed the next generation mechanical transmitter as the outside diameter of the magnet is varied

this is shown only to elaborate on the general trend. At the lower bound, the symbolic solution to Eq. 6.22 produced in MATLAB is shown by the dotted blue line.

Rightfully we find the position of the lower bound with respect to η_c in Fig. 6.9 is a little disappointing. By inspection it is clear this is in fact not the absolute maximum possible efficiency. The problem occurs because $P_{drive,wc}$ and $P_{rotor,wc}$ in Eq. 6.22 are both functions of τ_{req}. As a result of limited digits of precision in MATLAB and rounding errors due to differing precision of various input constants, the resulting sixth-order polynomial easily includes countless approximations. The precision was increased to 32 digits and all variables were replaced by a corresponding symbolic value until the last step of computation, which then results in the minimum OD as 0.9093 in shown as a vertical green dotted line in the figure.

Fortunately the numerical computation of torque for any arbitrary OD is computationally efficient.[4] Instead of investigating further the problem is solved numerically using the curve

[4] Identifying the OD to within 0.0001″ will suffice given manufacturing constraints.

of Fig. 6.9 where the maximum is found to be 0.8775 in, plotted as a vertical purple dashed line in the figure.

The boundary conditions for power consumption for operation in the entire target carrier frequency range are shown in the plot of Fig. 6.10. The red dashed line is once again OD_{max} and operation cannot extend beyond OD_{max} without replacing the motor. Corresponding magnetization field is shown on the right y axis (purple) for reference.

We can use the two y-axes as a visual aid for cost-benefit analysis. Operation at OD_{min} achieves maximum *field generation efficiency*. On the other hand the magnetization field at OD_{max} is a factor of $\sim 3\times$ greater than at OD_{min}. By minimizing the required number transmitter cells to generate a given $\vec{B}$ field, operation at OD_{max} achieves maximum *system scaling efficiency*.

Finally, it is important to evaluate the field generation efficiency across the operating region. Figure 6.11 shows the field generation efficiency across frequency. Unsurprisingly the efficiency improves when the operating frequency is reduced. The relationship between the boundaries shows another interesting effect, for small magnets the efficiency increases by $\sim 14\times$ from the lower to the upper boundary, where for large magnets this is only $\sim 6\times$. This suggests for magnets close to the minimum size there is a significant value to the use of lowered carrier frequency compared with large magnets where the effect is less lucrative.

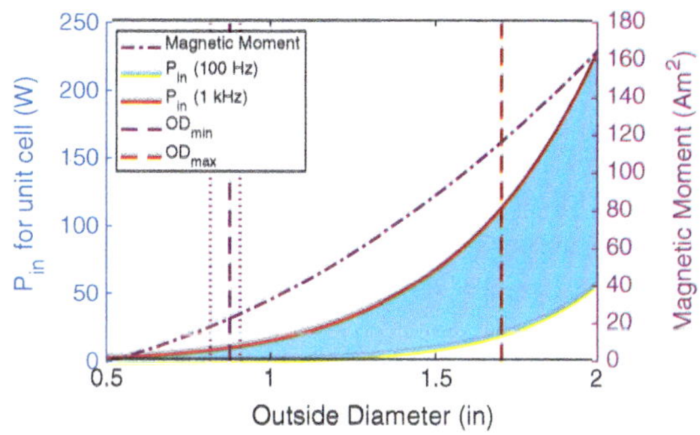

Fig. 6.10 Modulation power (at α_{max}) and magnetic moment for varying outside diameter of the magnet. Upper and lower bounds are shown at the edges of the channel (100 Hz and 1 kHz)

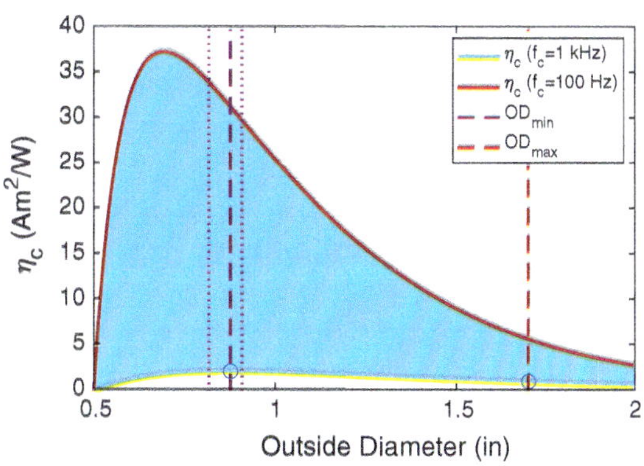

Fig. 6.11 A bounded region of operation in terms of field generation efficiency shown for the 100–1000 Hz channel. The position of optimization at 1 kHz found in Fig. 6.9 is circled

As an aside, $\eta_{c,max}$ is achieved with a smaller magnet for 100 Hz than for 1 kHz. This is unsurprising given $P \propto \omega(OD/2)^2$ and $m \propto (OD/2)^2$. Finally, note that our optima will change significantly if losses due to air friction are reduced by reducing the air pressure around the rotating magnet, i.e., by operating the transmitter in a partial vacuum.

6.4.2.1 Optimization at Cost, Remarks

Presented with a cylindrical magnet having fixed inside diameter, fixed length, and variable outside diameter, the model identifies the position of two conceptual optima. We will continue to explore both conditions and resulting impact to system performance as a reference case in further design effort.

It is worthwhile to note that these power efficiency models are all presented in the worst case. Regardless of the specifics in hardware design, the model simulates the absolute maximum power consumption and absolute minimum efficiency which result. It is likely possible to achieve better performance through careful design of efficient encoding and modulation schemes which do not make use of the full capabilities as specified.

Operating at OD$_{max}$ maximizes the field generation of each cell. The sensitivity of magnetic induction in the receiver increases with frequency, motivating the use of high carrier frequencies in the channel, where the improved efficiency for reduced magnet size is not as large. Under these guidelines a large magnet specified close to the maximum allowable size to achieve maximum field generation per cell. A magnet with OD 1.625 inches (41.28 mm) and length 3 inches (76.2 mm) has a $\tau_{req,wc}$ of 13.4 mNm in the worst case (maximum air friction, 1000 Hz). This is not a complete picture as $\tau_{req,wc}$ is for peak modulation rate and the average modulation torque $\tau_{mod} < \tau_{req,wc}$ will be much lower. Regardless the assembly will operate at a nominal carrier frequency below the limit of the channel which reduces τ_{af} and the motor is rated to handle transients well above the maximum continuous rating.

Operating at OD$_{min}$ maximizes field generation efficiency of each unit cell and minimizes inertia. Even though many more parallel subsystems require increased complexity in the controller, this is an economy of scale. It is preferable to work in the lower limits of the channel as the skin depth is reduced and by minimizing inertia there is greater overhead to gain performance by enhancing the modulation scheme. A magnet with OD 0.75 inches (19.05 mm) and length 3 inches (76.2 mm) has a $\tau_{req,wc}$ of 1.3 mNm in the worst case (maximum air friction, 1000 Hz). The dramatic change from OD$_{max}$ is in great part due to the reduction of air friction around the surface.

6.5 Modular Transmitter Design

6.5.1 Resolving the Field Generation Limit

In Sect. 6.2.3.1 we avoid treating field generation as an absolute parameter and instead opt to consider a scalable structure. We now revisit the subject. Effective transmit distance is set by

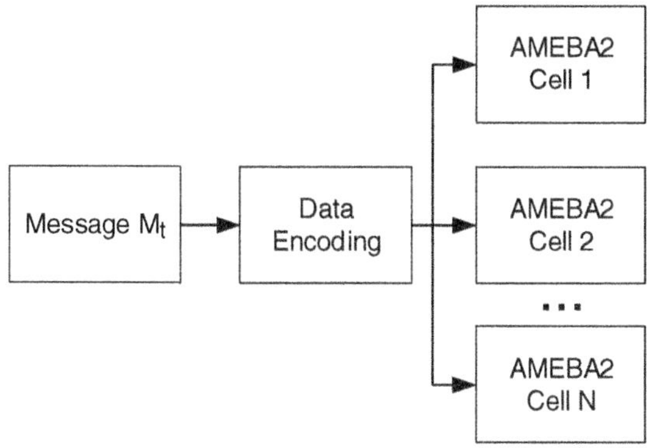

Fig. 6.12 An outline of the principal concept of multi-cell mechanically based transmitters. The same signal m_i is transmitted by each of N unit cells

the total volume of magnetic material. In a single transmitter, increasing the generated field increases the total torque requirement. Since the constrained system is limited fundamentally by the available torque of an electric motor at the desired carrier frequency, the effective transmit distance is fixed. That is, the effective transmit distance would be fixed if the transmitter were comprised solely of a single element. A parallel array of elements, or *unit cells*, each operating as a constrained fundamental model satisfying Eq. 6.1, could certainly be scaled to any desired volume of magnetic material to meet the specified transmit distance.

Such a system would operate as multiple unit cells working in parallel. As in the block diagram of Fig. 6.12, the message M_t is encoded as for a single unit cell, then transmitted using multiple unit cells simultaneously. If all cells have individual magnetic moments given by $m = \frac{1}{mu_0} B_r V$, the result is a single transmit signal effectively produced by a source with magnetic moment $m = \frac{N}{mu_0} B_r V$.

Introducing multiple unit cells creates an additional control system requirement. If the transmitted signals produced by each unit cell are not in phase, the net magnetic moment produced by the assembly will be $< N$. To compensate, a phase controller is required. An incremental encoder producing one tick per full revolution of the rotor is installed in the motor of each unit cell transmitter.

The system layout with addition of a phase controller is shown in Fig. 6.13 and allows the distributed transmitter elements to be synchronized with one another. The design operates on the same principle as a phase-locked loop (PLL). One unit cell is defined as the master, having phase 0° at the rising edge of the incremental encoder. Remaining cells are synchronized by adjustment to the feed forward control signal based on the phase of each with reference to the master. This adds another pole to the control system which must act to maintain phase synchronization between the elements without disturbing the structure of the transmitted signal.

The accuracy of the phase controller is then limited by manufacturing capabilities, requiring that the dipole moment of the magnet in each unit cell is aligned with the single-tick-per-rotation encoder in an identical fashion for each element. If the elements are not perfectly aligned after assembly of each cell, a second encoder with a large number of ticks per rev-

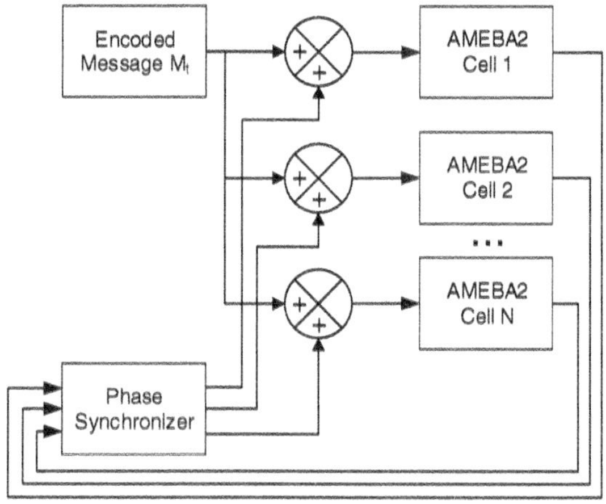

Fig. 6.13 A phase synchronizer ensures all cells oscillate in phase and contribute constructively to the total field generation

olution could be used to add an error offset. The error margin can be determined for each cell individually by running a simple experiment.

Using multiple unit cells in parallel allows for additional features of particular interest to enhancing the robustness and capabilities of the AMEBA2 system as a whole. One of the principal goals for the original work is to produce a power efficient system. Given the models for effective transmit distance and BER based on the strength of the generated field that have already been established in Chap. 5, the number of active unit cells can be configured based on the distance between transmitter and receiver. Depending on the actual location of use and the medium to be transmitted through, the channel will have different characteristics which the system can adapt to by enabling additional cells to increase the strength of the transmitted signal. In addition the system can include one or more cold spares which can be enabled in the event one of the unit cells has failed to operate, increasing the reliability of the transmitter in an emergency situation where such a failure could be otherwise detrimental.

Multiple unit cells also allows the use of multi-carrier modulation, the transmission of multiple signals in parallel. One method to achieve this is orthogonal frequency division multiplexing (OFDM) [8]. In an OFDM communication channel, multiple carriers are spaced orthogonally to one another and each carrier transmits a different data stream. The requirement to maintain orthogonality is the same as for orthogonality of the symbols in one carrier for non-coherent detection, the space between each carrier in the frequency domain must be $\Delta f > R_s$. Figure 6.14 shows one approach where the message M_t is divided into two sections with symbols m_i and m_j, each portion transmitted by two of four cells simultaneously.

Considering again the case of a multi-cell AMEBA2 transmitter with $N = 4$, the robustness of cold-spare unit cells to adjust energy consumption and overall field generation to adapt to the conditions of the channel can be retained. When the channel is of good quality and the transmitter is close enough to the receiver, the second cell for each portion m_i and m_j of the message in Fig. 6.14 can be switched off. Additional units can also be implemented as

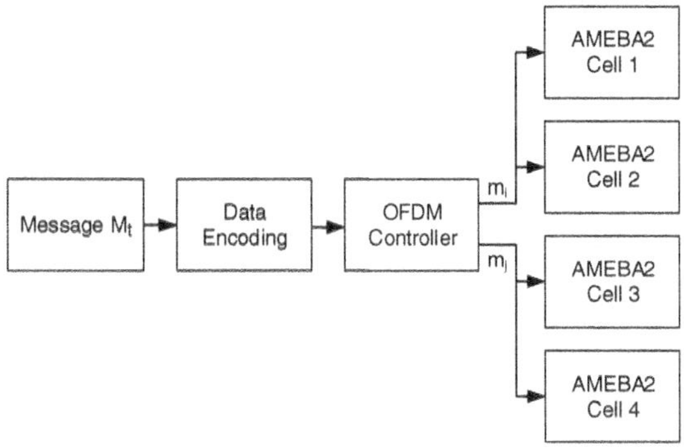

Fig. 6.14 Multi-carrier modulation in an $N = 4$ multi-cell AMEBA2 transmitter is quite straightforward by addition of an OFDM block. In the figure two sub-messages (m_i and m_j) are transmitted simultaneously with two unit cells per carrier

cold spares which may be activated as needed. The conditions of the channel at each carrier frequency may be different, and it would also be possible to change the number of cells devoted to each set of symbols based on the conditions of the channel seen by each carrier.

Typical applications of OFDM use a large number of carriers spaced as close together as is possible to achieve high bandwidth while maintaining spectral efficiency [8]. As a consequence the receiver must be accurately tuned to the transmitter frequency to avoid loss of orthogonality (i.e., crosstalk between carriers as observed on the receive side). In the AMEBA2 system, the use of CP-FSK makes it difficult to constrain the bandwidth of any given carrier as instantaneous frequency of the transmitted signal tends to walk (see Fig. 5.24). A large bandwidth for a given carrier reduces the required efficiency of the data encoder and increases robustness against challenges with extremely short messages being composed of non-random sets of characters, even after applying a compression algorithm. This in turn implies also implies a decrease in overall power consumption when the channels have low spectral efficiency. Fortunately the unusual choice of 100–1000 Hz transmission band makes the concern of spectral efficiency less critical.

Given the considerations to possible benefits of OFDM, the proposed system for the realized AMEBA2 transmitter includes the use of four unit cells as shown in Fig. 6.15, with the target goal being to transmit data on up to three carriers simultaneously. The fourth cell will be used as a cold spare to increase the strength of the generated field for the channel with the greatest attenuation. In the event any number of cells is compromised, the system will adapt by changing the message encoding to use fewer carriers and devote more cells to the remaining carriers. The block diagram of Fig. 6.15 shows only four cells, however the use of modular design allows for the implementation of any number of additional cells as desired.

Phase synchronization in the implementation of a multiple-cell multiple-carrier system with OFDM can be achieved in the same fashion as for the case of a basic multi-cell layout. Cells transmitting the same signal absolutely must have phase synchronization. Signals in different carrier channels also require phase synchronization. Each carrier is ideally offset

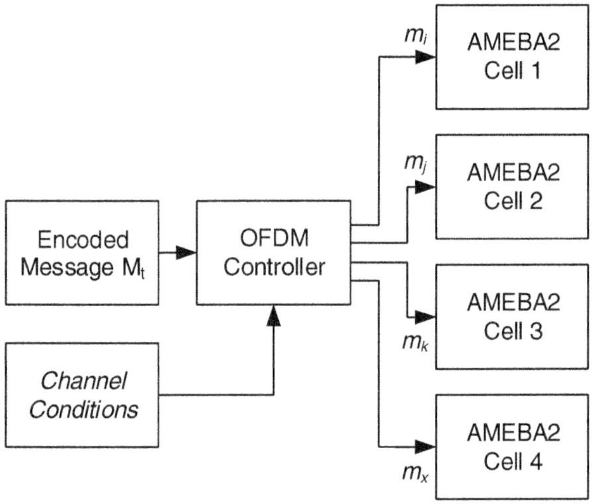

Fig. 6.15 Implementating multi-carrier modulation in an $N = 4$ multi-cell AMEBA2 transmitter is quite straightforward by addition of an OFDM controller in the digital control system. In this case, two sub-messages with symbols m_i and m_j are transmitted on two unit cells per carrier

in phase for efficient detection, though by maintaining orthogonality between the carriers it would not be destructive if they were not.

6.5.2 BER Analysis of Optimum Magnet Design

The field strength and BER for 1–4 unit cells is now analyzed in comparison with the prototype. This analysis is performed using the same model as is used for the prototype in Sect. 4.3.5, with minor improvements. The approach is revised to consider the channel conditions (discussed here) independent from the performance of the receiver (discussed in Chap. 9). Field strength used for analysis is the maximum field strength available at the position of the receiver, given by Eq. 3.2 for the 1-axis case and Eq. 3.3 for the 3-axis case. In all discussion which follows in this section, field strength is for a 1-axis receiver unless otherwise specified.

Field strength projections shown in Fig. 6.16 are in units of expected field strength. The AMEBA2 Unit Cell curve is the base case using a single transmitter cell and a one-axis sensitive receiver as was used in the prototype receiver. The predicted field strength was observed to be accurate in comparison with the experimentally determined trend in Fig. 5.11. Additional curves show the improvement in field strength detected when a 3-axis sensitive receiver and/or an additional unit cell is used. In comparison with the prototype described in Chap. 5, the field strength of AMEBA2 is more than an order of magnitude greater even in the base case.

The expected P_{be} for various $\frac{E_b}{N_0}$ computed for the prototype in Chap. 5 uses a model which assumed a rather optimistic improvement in the minimum Euclidean distance d_{min}^2 (not to be confused with the maximum effective transmit distance d_{max} used later in the chapter)

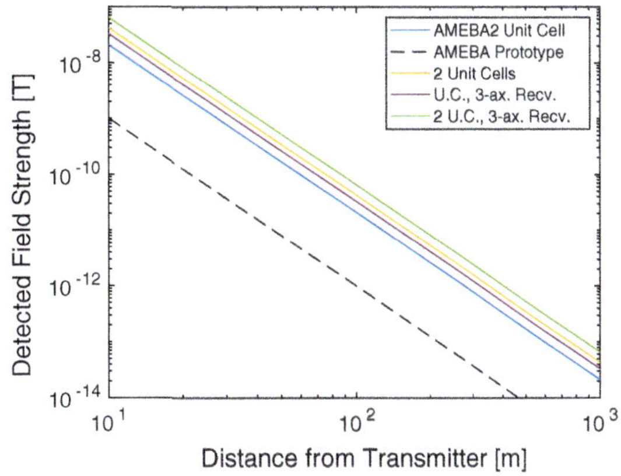

Fig. 6.16 Predicted field strength observed by the receiver using a 1-axis (1-ax) or 3-axis (3-ax) sensitive antenna. Curves are shown for various configurations of AMEBA2 unit cells (U.C.) in comparison with the prototype described in Chap. 5

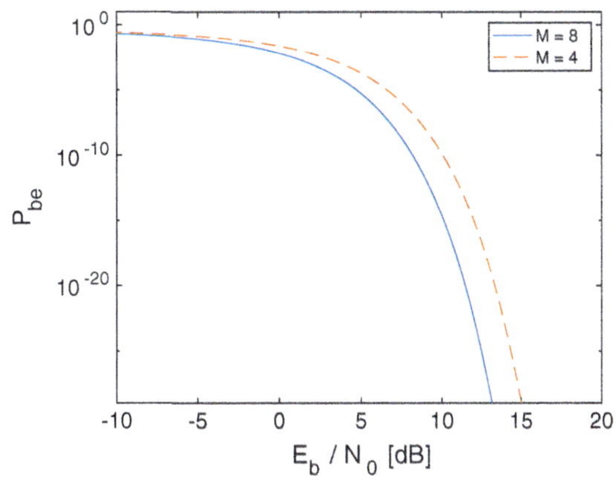

Fig. 6.17 Theoretical BER performance of a CP-FSK receiver in an AWGN channel. The model now assumes a tight spectral bandwidth based on the minimized Δf specified for AMEBA2

between symbols which is no longer accurate for the carefully optimized AMEBA2. For $T_s = 4$ the minimum required Δf is 0.25 Hz, but 4 Hz was used to overcome frequency instability in the transmitter and the limited resolution of the detector. Consequently d^2_{min} is increased and BER projections for the prototype include a small increase in d^2_{min} to account for this change. However, the adjustment is too small to account for the severe frequency instability issues in the prototype which results in the potential of overlap between symbols, decreasing the effective d^2_{min} despite the large spectral distance between symbols in the encoded CPM signal. Based on experimental observation, the prediction used for the prototype (shown in Fig. 4.12) is considered to be accurate.

Using a closed loop control system, the phase of each transmitted symbol is expected to closely follow the intended transmission signal and so Δf will now be the minimum value to minimize the transmit power P_s per symbol ($P_s \propto m_i \Delta f$). The result is a tight spectral bandwidth and the new model of Fig. 6.17 uses a realistic value for d^2_{min} to account for this case, using the value of d^2_{min} given in [23] for $h = 1$.

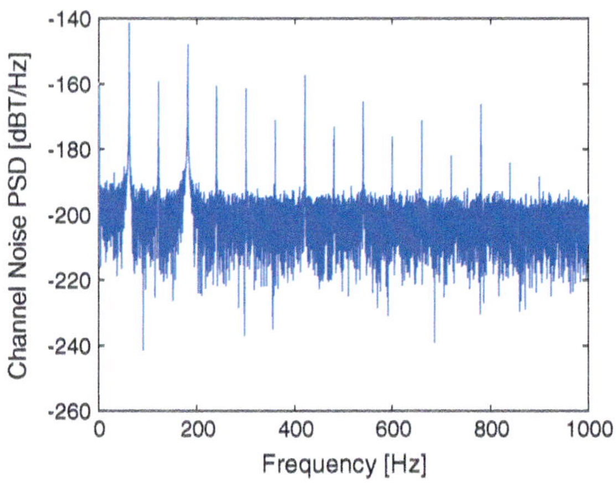

Fig. 6.18 Experimentally observed conditions of the ELF channel. Data collected on 2020-10-01 in our laboratory while preparing to take measurements of Sect. 4.2. The receiver is spaced ∼9 ft from the nearest known source of 60 Hz noise (a power line in a ceiling light enclosure)

Projected error performance is dependent on the conditions of the ELF channel. For design analysis, a noise PSD is determined based on data collected with the receiver outside the door of our laboratory. This is presented in Fig. 6.18 which shows a noise PSD for which the AWGN portions of the noise are between –220 dBT/Hz and –190 dBT/Hz. The spectrum includes the 0–100 Hz region which may be used for start-of-transmission and end-of-transmission control sequences.

60 Hz harmonics are present across the band. Despite their decreasing magnitude toward the upper end of the channel, these harmonics are still quite strong and remain boundaries which cannot be crossed by the transmit waveform, since the noise is far too high which increases P_{be} dramatically. Furthermore these sources of noise must be removed in the receiver before input to the preamplifier to avoid saturation so any transmit signal at these frequencies would not be detectable for both reasons given. The channel is effectively split into sub-channels by the presence of these harmonics before any consideration is made as to the selection of sub-channels for single- or multi-carrier modulation.

For AMEBA2, BER analysis is now considered under the conditions of the channel alone, the sensitivity and noise floor of the receiver antenna are considered separately as signal and noise are both amplified equally by the receiver. Under the experimentally observed conditions of the channel in our laboratory, Fig. 6.19 shows projected BER performance over distance in various cases. Vertical dashed light-red lines show the intersection of each characteristic curve with the point where $P_{be} = 10^{-3}$. In the base case (a single AMEBA2 unit cell with a 1-axis receiver), the expected maximum transmit distance d_{max} is 35.1 m.

The reference case for the prototype considers P_{be} if the magnet in AMEBA2 were replaced with a magnet of the same volume and grade of NdFeB material. All other parameters are identical (i.e., this curve does not reflect the projected BER for the complete prototype transmitter).

Table 6.2 gives the maximum transmit distance predicted in the base case compared with the equivalent prototype and configurations which use additional cells and/or a

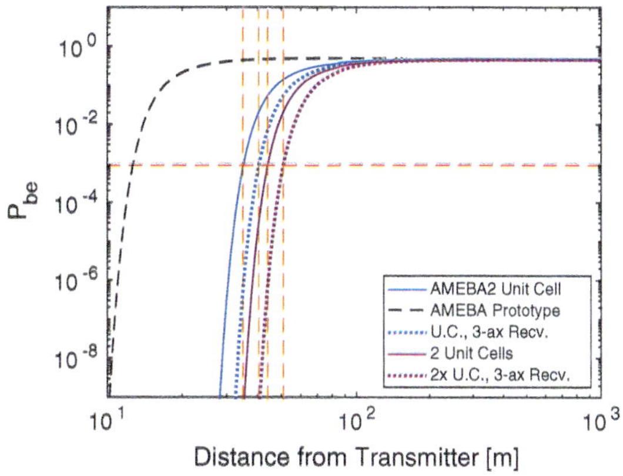

Fig. 6.19 Projected BER performance over distance under experimentally sampled channel conditions for various configurations of transmitter cells using (i) a 1-axis (1-ax) receiver and (ii) a 3-axis sensitive (3-ax) receiver. The horizontal line denotes $P_{be} = 10^{-3}$

Table 6.2 Maximum transmit distance based on conditions of the ELF channel for various configurations of the transmitter using the noise PSD of Fig. 6.18

Transmitter configuration	Maximum transmit distance (d_{max}) (m)
AMEBA prototype	12.8
AMEBA2 unit cell (Base case)	35.1
2x Unit cells	44.3
4x Unit cells	55.8
1x Unit cell (3-ax receiver)	40.7
2x Unit cells (3-axis receiver)	51.2
4x Unit cells (3-axis receiver)	64.5

3-axis sensitive receiver. Doubling the number of cells (1→2, 2→4) increases the maximum transmit distance by 26%. For any given case using a 3-axis sensitive receiver increases the effective transmit distance by 16%.

There are diminishing returns from increasing E_b alone. Decreasing the noise N_0 at the receiver is another method to improve system performance. Noise power within a sub-channel may be reduced by oversampling, which reduces the power of uncorrelated noise sources (such as AWGN) by a factor of N where $N = f_s/f_n$.[5] Computer simulation of such noise reduction using the same dataset is shown in Fig. 6.20 assuming a sample rate of 1 MSa/s. Correlated noise is not reduced through this approach and as 60 Hz harmonics are not removed, the orange curve of reduced noise PSD in the figure is disjoint as indicated by the blue dashed lines. Notch filtering is performed in software to 60 Hz harmonics for clarity.

A projection of BER over distance using reduced noise power by oversampling is shown in Fig. 6.21. The situation is now more promising, maximum transmit distance in the base

[5] Here f_n refers to the Nyquist rate or $2f_{max}$.

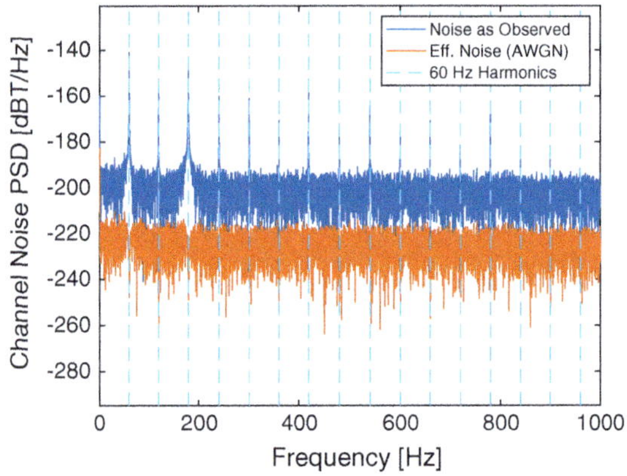

Fig. 6.20 Experimentally observed conditions of the ELF channel and the effective noise power when data is oversampled. Digital notch filtering is used to 60 Hz harmonics in the figure for visualization purposes. Data collected on 2020-10-01 in our laboratory, with the receiver spaced ∼9 ft from the nearest known source of 60 Hz noise (a power line in a ceiling light enclosure)

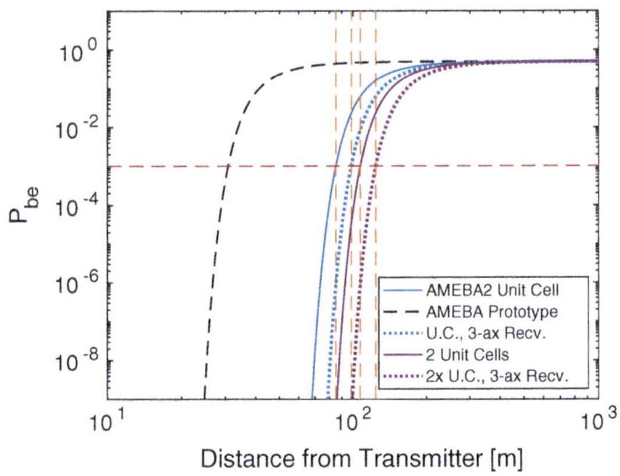

Fig. 6.21 Projected BER performance over distance under experimentally sampled channel conditions for various configurations of transmitter cells using (i) a 1-axis (1-ax) receiver and (ii) a 3-axis sensitive (3-ax) receiver. The horizontal line denotes $P_{be} = 10^{-3}$

case is now 84.9 m. However it is clear to reach the 1000 m goal that E_b/N_0 must be further improved. E_b can always be increased by adding additional transmitters. As for N_0, it is well known in the field that channel conditions vary by location. Beyond 60 Hz harmonics, it is not yet known what a typical magnetic field noise PSD is at ELF frequencies.

Tabulated values for d_{max} accounting for the effect of oversampling are shown in Table 6.3. The trends in % improvement are the same as for the data in Table 6.2, however the result is more dramatic as the base case d_{max} is larger. This is a rather positive result as it indicates the benefits to multiple unit cells or a 3-axis receiver are greater when the conditions of the channel are improved.

Before further considerations on improving the maximum transmit distance are made, it is prudent to study the conditions of the channel in various settings and consider the effects

Table 6.3 Maximum transmit distance based on conditions of the ELF channel for various configurations of the transmitter using the noise PSD with simulated oversampling shown in Fig. 6.20

Transmitter configuration	Maximum transmit distance (d_{max}) (m)
AMEBA prototype	30.9
AMEBA2 unit cell (Base case)	84.9
2x Unit cells	107.0
4x Unit cells	134.8
1x Unit cell (3-ax receiver)	98.3
2x Unit cells (3-axis receiver)	123.8
4x Unit cells (3-axis receiver)	156.0

of the environment surrounding the receiver. The results of our investigation into N_0 were already described in Sect. 4.2.

Other approaches to increasing d_{max} such as increasing symbol energy by increasing T_s or increasing Δf to increase d_{min}^2 are also possible; these effects cannot be considered until the complete receiver is understood. Discussion on effective transmit distance continues in Chap. 9.

References

1. Glickstein, J. S., Liang, J., Choi, S., Madanayake, A., & Mandal, S. (2019). Power-efficient ELF wireless communications using electro-mechanical transmitters. *IEEE Access, 8*, 2455–2471.
2. Hughes, A., & Drury, B. (2019). *Electric motors and drives: Fundamentals, types and applications*. Newnes.
3. Chao, W. Y., Xin, H. Y., Wei, W., & Rong, L. J. (2017). Frictional analysis of deep-groove ball bearings with varying circumference radial preloads. *Advances in Mechanical Engineering, 9*(5), 1687814017703895.
4. Ricci, M. C. (2015). Ball's motion, sliding friction, and internal load distribution in a high-speed ball bearing subjected to a combined radial, thrust, and moment load, applied to the inner ring's center of mass: Mathematical model. *Journal of Physics: Conference Series, 641*(1), 012017. IOP Publishing.
5. White, F. M. (1998). Fluid Mechanics (4th Ed.), New York: McGraw-Hill.
6. White, F. M. (1974). *Viscous Fluid Flow*. New York: McGraw-Hill.
7. Papanikolaou, M., Frank, M., & Drikakis, D. (2017). Effects of surface roughness on shear viscosity. *Physical Review E, 95*(3), 033108.
8. Silver, H. W. (2020). The ARRL handbook for radio communications. The National Association for Amateur Radio.

7 Implementation of the Closed-Loop Transmitter

7.1 Mechanical Design

There exist some parameters which have no defined ideal value—these include system complexity, manufacturing time, cost, level of redundancy, and diversity of available parallelized features (see Sect. 6.5.1). As such, much like the true BER of a CP-FSK modulation scheme, it would be impossible to calculate an exact specification for optimum efficiency. Hence (concerning the design of the unit cell), we prepare to explore the boundary conditions identified in Sect. 6.3.4.

At the position of OD_{max}, we propose to maximize the rotational speed of the assembly and operate within a higher frequency region of the channel. Here, we expect

- Increased $|B|$ induced in the receiver,
- Reduced interference 60 Hz and higher harmonics in the channel, and
- Decreased flicker noise in the receiver

where we further aim to maximize the total field generation by the transmitter by maximizing the size of the magnet.

We also consider operation at the position of OD_{min} by approaching the problem of field generation as one which requires power efficiency above all else.

7.1.1 Selecting a Motor

At a particular maximum speed of interest, maximum torque is a direct result of design efficiency for the electric motor. Torque $\tau = kI$ where k (the motor constant set by the design of the motor) and I are both limited by electrical efficiency from winding resistance

J. S. Glickstein and S. Mandal, *Electromechanical Transmitters for ELF/VLF Radio*,
Synthesis Lectures on Engineering, Science, and Technology,
https://doi.org/10.1007/978-3-031-15123-1_7

and configuration. A thorough review of available motors on the market indicates these parameters are typically about the same for any given manufacturer of high-speed power-efficient motors. On the other hand, mechanical efficiency is based on the size of the motor as some of the torque produced by the motor is required to drive the shaft of the motor itself ($\tau_{output} = \tau_{max} - \tau_{rotor}$) and power loss is greater for larger armature wiring. This factor motivates the use of a high-speed electric motor which is both small and electrically efficient.

Another problem with the AMEBA prototype which has not yet been addressed is the frequency instability of the generated signal. The open-loop control system did not accurately maintain a constant motor speed. Speed variations could be as high as $\pm 0.5\Delta f$, making reliable decoding of the transmitted signal difficult. Frequency variance in the transmitted signal increases the likelihood of errors in the detector, well beyond the error rate predicted by the models in Sect. 4.3.5. In addition, the dominant limit to the symbol rate in the prototype was a result of the required increase in frequency domain resolution to compensate for the effects of this variance. AMEBA2 is designed to be robust against this problem through the use of a carefully controlled closed-loop feedback system. The motor must be compatible with an encoder which provides sufficient resolution to accurately control the speed of the motor.

7.1.1.1 The Contenders

A Swiss company specializing in high-precision microdrives, maxon motor AG, offers a large selection of highly suitable candidates. The ECX series of high-speed motors includes offerings with continuous operation at speeds beyond 60 krpm and can be mated to a precision encoder. One particular selection (maxon ECX SP16L) has a no-load speed of 64.6 krpm, corresponding to a frequency of 1076 Hz. This provides a reasonable operating bandwidth even in a case where the target carrier frequency is set to the maximum 1000 Hz within the range of proposed ELF transmission frequencies.

On the other hand, we can gain greater torque at a comparable physical scale by reducing the target frequency to 100–200 Hz. Switching to a motor with a larger body and a four-pole rotor, we consider a maxon EC-i 40 BLDC motor having a no-load speed of 13.2 krpm (220 Hz).

From the available options at this rated speed, the next design parameter is torque. The final selection is the motor which provides the maximum continuous torque rating, an ECX SPEED 16 L Ø 16 (shown in Fig. 7.1) with armature wiring rated for 24 V and a maximum efficiency of 89.5%. This motor is quite small, and as a result can operate at high speeds as required. It is also electrically efficient and can deliver near maximum torque (maximum being the theoretical torque delivered by a textbook motor having zero armature inertia and zero armature resistance). The manufacturer can customize this motor to suit the intended application.

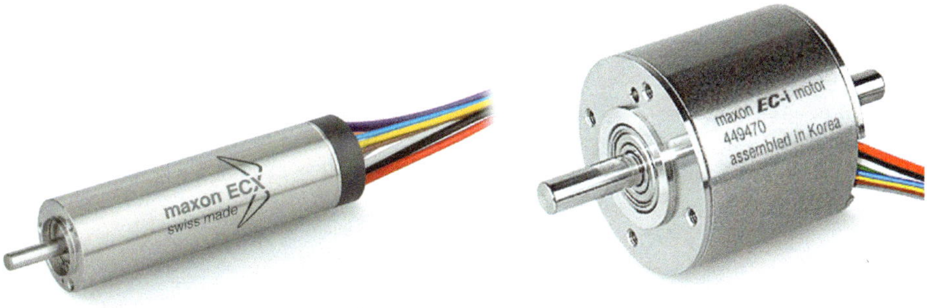

Fig. 7.1 Two contenders, maxon ECX SP16L (left) and EC-i 40 (right) precision microdrives

Rated voltage for the armature wiring may be selected as 12 or 24, 36, and 48 V. During a consultation with the vendor, we learn this selection is intended to be compatible with the desired motor controller. However, the resulting thermal properties of each armature are slightly different which results in a variation of the motor's torque constant k [2]. We are not restricted to any particular supply voltage and are free to choose based on the relative value of

- Marginal (5%) increase in torque rating at some values;
- Increased drive efficiency by reduction of motor current for larger V_{nom};
- Portability in the field for larger V_{nom};
- Magnitude of potential transients coupled through the power amplifier to the bridge driver.

A value of 24 V is selected as this yields the maximum torque compared with other options and maintains a good balance between drive current and back emf.

7.1.1.2 Specifications of the Motor

General specifications of the motor are tabulated in Table 7.1. The rated continuous torque is of particular interest as at first glance 16 mNm appears quite small. In the early stages of design, the approach here was to first select the desired field strength and calculate the required torque to drive the magnet, then the search for a motor was dominated by the desire for far greater torque at the operating speed. This approach is unreasonable given there is no motor which can deliver both high speed and high torque—recall the trade-offs between torque and speed due to the physical properties of the armature wiring described in Chap. 6. Instead, the magnet will be sized to fit the available torque.

Of course, it *might* be possible to increase the torque output by lowering the armature winding resistance, which could be accomplished by cooling the motor (e.g., by submerging it in liquid nitrogen). This has been well established as a formidable challenge and a parallelized array of distributed magnetic elements is far more practical.

Table 7.1 General specifications of two motors selected for speed and torque, respectively

Specification	Motor rating	
	ECX SP16L	EC-i 40
Nominal voltage (V)	24	24
n_{max} (Continuous) (rpm)	64 600	13 400
τ_{max} (Continuous) (mNm)	16	52.8
I_{max} (A)	4.82	2.8
τ_{stall} (mNm)	407	810
I_{stall} (A)	115	47.9
η_{elec} (%)	89.5	85

Table 7.2 Motor characteristics relevant for design of the motor driver and control system. Here, τ_{mech} is the response of the motor loaded only by the inertia of the armature. τ_{elec} was not provided directly by the manufacturer; it has been calculated from the provided coil parameters

Specification	Motor rating	
	ECX SP16L	EC-i 40
Coil resistance (Phase to Phase) (Ω)	0.208	0.501
Coil inductance (Phase to Phase)	14.1 μH	0.39 mH
Torque constant (mNm/A)	3.53	16.9
Speed constant (rpm/V)	2700	565
τ_{mech} (No load) (ms)	2	1.84
τ_{elec}	67.8 ns	77.8 μs
Rotor inertia (gcm^2)	1.2	10.5

The motor is a three-phase electronically commutated brushless DC motor (3Φ EC BLDC), requiring three half-bridge drivers and a feedback system to control. The three-phase armature wiring yields higher efficiency and lower τ_{elec} as a result of three smaller armature coils replacing one large one in a comparable single-phase approach. This also gives greater control over the speed as the motor can respond more quickly and thus the feed-forward loop bandwidth may be greater for the control system. Relevant electromechanical characteristics of the motor required to complete the design of the unit cell are shown in Table 7.2 for the interested reader. Greater detail is given in the discussion of the motor controller.

Regarding the encoder, a speed feedback encoder is selected with the maximum number of ticks per revolution (counts per turn, cpt), 1024 cpt. The rotating encoder may be modeled as a variable frequency oscillator as a feedback input to the control system. In this case,

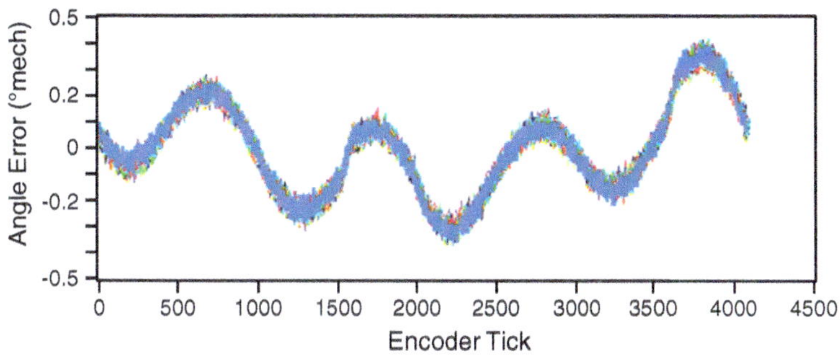

Fig. 7.2 Manufacturer-provided plot of feedback angle error from experimentally determined data for a 1024 cpt quadrature encoder. Data collected with the EASYENX encoder supplied with $V_{CC} = 5$ V and $n = 83.3$ rps at 25 °C. The X-axis is an incremental count of encoder ticks over one full revolution of the armature. 25 revolutions were measured with results overlaid. Angle error is shown as an accumulated error compared to a 0° reference from an absolute encoder (1 cpt)

manufacturer specifications of angle error and jitter correlate to control system parameters describing the error of the system. For the case of continuous speed operation,

$$v(t) = V_0 \mathrm{sgn}(\sin(\omega_0 t + \phi(t))) + e(t)$$

where $\phi(t)$ is the phase error which for a maxon EASYENX encoder is at most 0.1° [1]. Using a digital detector, the amplitude error $e(t)$ may be neglected.

The encoder chosen is the maxon EASYENX encoder. Manufacturer-provided angle error per turn is shown in Fig. 7.2. The X-axis of the plot is in units of quadrature ticks over a single revolution. The EASYENX encoder is in a standard quadrature configuration with two 1024 cpt encoders spaced 90° out of phase, so the actual resolution of the encoder feedback is 4096 cpt for an edge-triggered receiver (12 bits).

Table 7.3 gives tabulated data from the experimental analysis as provided by the manufacturer. Jitter gives the maximum position error in degrees per full revolution of the motor as averaged over the specified number of full rotations, in this case 25. This sets the maximum position repeatability when the present position is calculated from a rolling average. INL is the peak-to-peak value of armature angle error over a complete rotation, which gives the maximum angle error for any single instantaneous position measurement (the peak-to-peak value of angle error in Fig. 7.2). This is the average deviation of armature position from the expected value. These two values define the position control accuracy in the mechanical domain.

DNL is the maximum average state error, the maximum average deviation of encoder state change from the expected value. One state is defined as the nominal space between encoder edges in the time domain, so DNL is the average deviation in terms of the expected duration of each nominal state between edges on either encoder. The minimum and maximum states are the instantaneous minimum and maximum error of each state given in units of bits, and

Table 7.3 Characteristic performance of the EASYENX encoder with 1024 cpt

Parameter	Measurement
Jitter	0.1°m (≈1 LSB)
Integral nonlinearity (INL)	0.7°m
Differential nonlinearity (DNL)	0.5 LSB
Min state	0.8 LSB (72°e)
Max state	1.5 LSB (135°e)

in degrees referenced to one complete pulse on either channel of the encoder. These values define position control accuracy in the electrical domain.

More details on the use of the encoder are discussed in Sect. 8.2.2.

7.1.2 Air Supply

Air bushings operate by producing a thin film of air between the bushing and a rotating shaft. A minimum air quality comparable to ISO 8573.1 Class 4 must be maintained for proper (read: frictionless) operation. Specific recommendations are

1. Dirt particle size $<15\ \mu m$;
2. Dew point $<-20°$ F (–4° F);
3. Oil/vapor content $<1\ mg/m^3$.

This is one of the finer points in tuning the assembly, requiring a pneumatic system to supply dry air free of dirt and oil to an acceptable level. Without it, contaminant particles larger than the air gap in the bushing contact the surfaces and produce friction. Oils and water vapor cause similar issues, though they also tend to accumulate, i.e., their effects have long time constants over the span of several minutes to an hour during transmitter operation.

Pneumatic circuits are used to describe the flow of pressurized air in a manner analogous to the flow of current through electric circuits. The pneumatic circuit of Fig. 7.3 is one such case and shows the components of the air compressor and several stages of filtering which supply air to the bushings. Filtration and drying use two stages, a combination particle and water filter in series with an oil vapor filter. The supply does not adequately suppress the dew point below the maximum acceptable level for continuous operation (over a span of several days) but the test bench is not needed. Instead, an expansion coil consisting of 5 m high-pressure rubber pneumatic tubing is used between the compressor output and filtration input to trap excess condensate, which must be removed and drained weekly.

A secondary pressure relieving valve after filtration is used to prevent downstream pressure in excess of the maximum rating for barbed tubing connections. Otherwise, it is possible

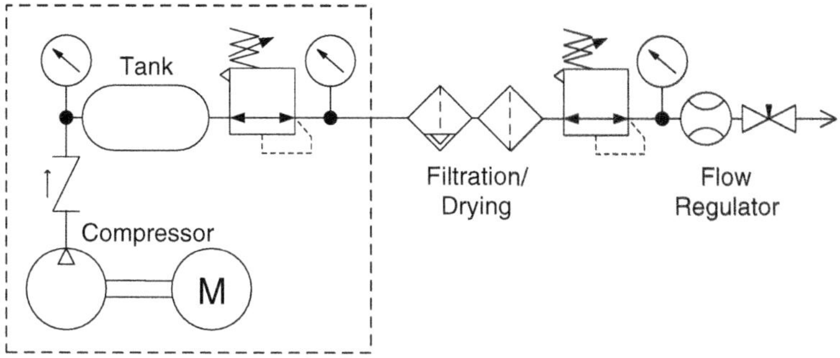

Fig. 7.3 Test bench air supply setup consisting of an air compressor and separate air pressure, flow, and quality conditioning required to produce an effective air film in the air bushings

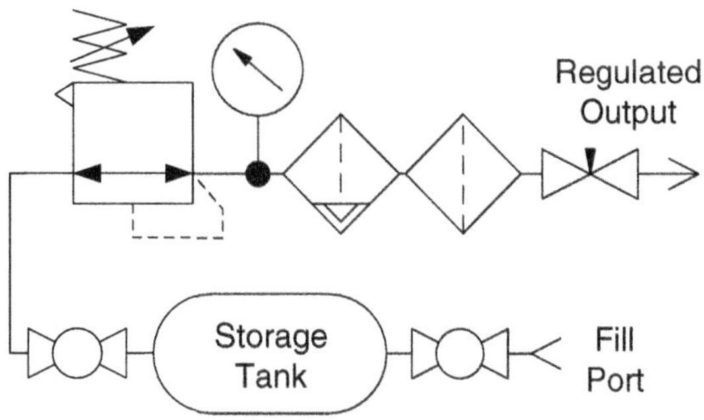

Fig. 7.4 Air supply of the mobile communication device where a storage tank replaces an air compressor similar to batteries replacing bench power supplies

for tubing to pop off a barbed fitting while the transmitter is rotating, causing a sudden spike in torque loading of the motor. A needle valve is used at the output for fine adjustment of flow and is shown in combination with a symbolic flow rate indicator, which is somewhat unhelpful in practice. Flow rate is preferably adjusted by feel, opening the needle valve slowly while the transmitter rotates at continuous speed and observing the current consumption of the motor, and listening for whirring noises characteristic of either bushing producing too much or too little lift. The flow meter remains useful as a reference for tank depletion rate.

In the field, a storage tank replaces the air compressor for mobile operation, which is shown in Fig. 7.4. The fill port at the input to the storage tank could be temporarily connected to a supply if one were readily available, but in a simulated disaster environment one must plan to be prepared. Output filtration remains the same and a needle valve is retained to regulate flow. The needle valve now serves as a set-and-forget calibration and the on-off switch for the supply is a ball valve at the output of the tank. The field air supply is used in Chap. 10.

A flow multiplexer is also useful for supplying air to many transmitters operating in unison. Our analysis has considered up to four transmitters in unison which can be supplied by an unregulated flow divider as in Fig. 7.5. The transmitters will be installed together in a

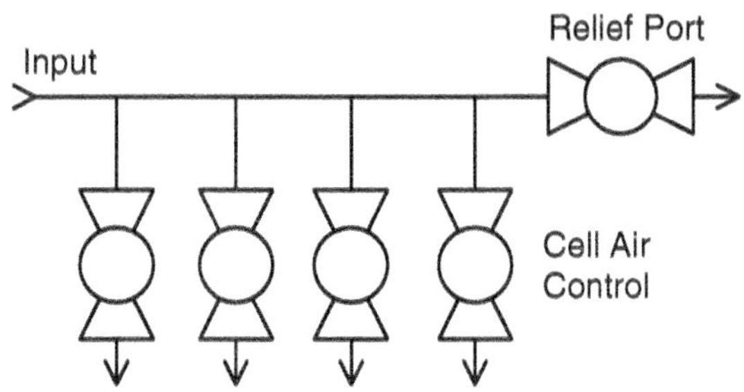

Fig. 7.5 Many transmitters in parallel are supplied with air through a flow multiplexer

single assembly and downstream pneumatic tubes will be short, however, each transmitter will have a variance in the required airflow. Optimally, the source pressure would be supplied above specification and regulated downstream as shown in Fig. 7.3, though wasted air due to pressure relief would increase by a factor of 4×, which is not suitable for field use. Instead, all mechanical cells can be supplied by a regulated air tank and the required flow to each unit is made adjustable using a needle valve. In this approach, there is a slight downside when the output pressure from the supply is below the set point of the regulator, in that changing the flow to any one unit can cause an upstream pressure drop which affects the flow to all other units. Adjustment to achieve optimal performance then becomes quite an art. Of course, the problem presents itself in either case, as we observed that the pressure and flow can be tricky to set for a single cell. This concept is not explored further for multi-cell configurations and is reserved for future work.

Copper water pipes are also used to run air supply lines in factories, machine shops, and even home workshops to power pneumatic tools as they are smaller, lighter, and cheaper than their electric counterparts. The flow divider is constructed from such pipes and ball valves from the local hardware store, and soldered using a propane blowtorch as if plumbing a water line. Quick-connect fittings are used to connect the flow divider to an air compressor in Fig. 7.6.

7.1.2.1 Lesser-Known Specifications of Pneumatic Hardware

Standards for copper plumbing pipe specify maximum static pressure for use in water supply, not air. Industrial mechanics who install such lines suggested that $\frac{1}{2}''$ type M copper pipe is acceptable below 150 psi (1 MPa) when soldered with lead-free solder, and de-rated to 100 psi if leaded solder is used. The recommendation was not to braze joints.

Low-impedance connections from a compressor or air tank to the regulator of each transmitter are made using $\frac{1}{4}''$ ID EPDM rubber air hose rated to 200 psi (2 MPa). PVC air hose is a low-cost alternative which is also less likely to kink in long runs, however, it is not recommended as it is too stiff to work within the field and does not completely uncoil.

After the air filter, we switch to $\frac{1}{8}''$ ID tubing. The tubing is firm polyurethane rubber having Shore durometer 95A. All used fittings are made of brass (preferred) otherwise

Fig. 7.6 Pneumatic flow divider attached to an air compressor

substituted with PVC meeting ASTM standard D1784 for strength and rigidity. There is a great variation in the actual dimensioning of fittings, and tubing from two manufacturers carrying the same durometer rating can also be different. It is recommended to anyone repeating this work that one measures the dimensions of the hose barbs being used and match the dimensions to a similar standard to find a starting point for the required tubing, and to expect a process of trial and error.

7.1.3 Permanent Magnet

The maximally sized magnet is manufactured by Stanford Magnets from grade N52 NdFeB and has ID 0.5 inches, OD 1.625 inches, and length 3 inches. The inside diameter is minimally sized for a magnet of this length.

The minimally sized magnet has ID 0.25 inches, OD 0.75 inches, and length 3 inches. No manufacturer was able to produce a magnet with these dimensions, so it is constructed from four smaller sized sub-units each with length 0.75 inches. The sub-units are assembled onto a precision shaft using epoxy rated for high shear strength, as shown in Fig. 7.7. Each magnet is affixed and given 24 h to cure before the subsequent unit is added.

Fig. 7.7 Assembly of long magnets from many small sub-units

7.1.4 Shaft Coupling

The motor has a shaft diameter of 3 mm and the shaft for the driven portion of the assembly is required to be a precision ground steel shaft with diameter $\phi\ 0.25^{+0.0000}_{-0.0007}$ inches in order to use the specified air bearings. The coupling must be made to couple the motor shaft with the diameter specified in SI units to the rotating shaft with the diameter specified in US customary units.

7.1.5 Computer-Aided Design of the Mechanical Assembly

With all analysis complete, the mechanical assembly is sketched and designed in Autodesk Inventor, a computer-aided design (CAD) program. This process began long before the complete analysis was finished, to aid in the design decisions which were made earlier in the chapter. As the assembly neared completion, analysis was completed and the assembly was finalized.

A render of the final unit cell design is shown in Fig. 7.8. The assembly has dimensions $2.56'' \times 2.25'' \times 8''$ (H×W×D) not including the protrusion of the motor, which adds an additional $1.1''$ depth. The assembly is compact for efficient utilization of space and to minimize air friction and the effects of loading. The magnet has ID $0.5''$, OD $1\ \frac{5}{8}''$, and length $3''$. The body of the motor has diameter $0.63''$ and length $2.2''$, partially contained within the mounting block.

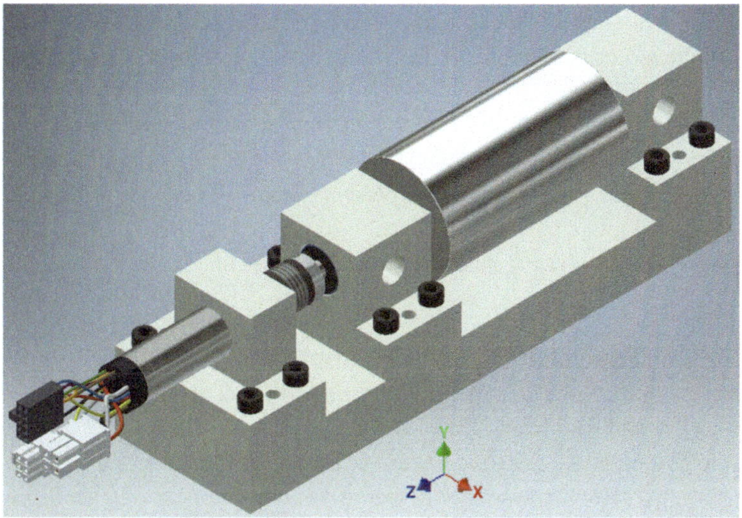

Fig. 7.8 An isometric render of the AMEBA2 unit cell design

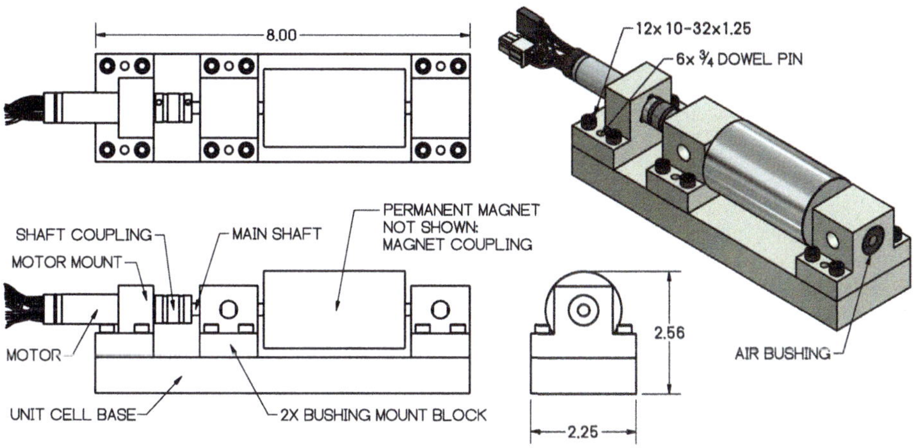

Fig. 7.9 Assembly drawing of the AMEBA2 unit cell. Dimensions are given in inches

The assembly drawing of Fig. 7.9 elucidates the complete configuration of the rotating assembly. Each part was modeled individually and then placed in an assembly where it is constrained to form the complete assembly. Constraint-based assembly design is useful to verify the fitment of parts, measure clearances, and ensure the shafts are both properly aligned and free to rotate. The shaft coupling appears off-center in the design, which is a remanent of design validation where the clearance between the shaft coupling and motor mount block has been measured when the shaft of the motor is fully extended into the coupling. This clearance was measured as 0.033 in. In the realized assembly the coupling is centered. The motor still extends far enough into the coupling to be fixtured properly in this case.

The motor and rotating shaft are fixtured by Delrin® mounting blocks aligned using dowel pins. The motor is intentionally exposed to allow thermal exchange with the air. A fan could be used if needed though the design is at present intended to operate within the rating for continuous operation in ambient air. The magnet coupling referred to in the drawing is not shown; it is an aluminum collar machined to a light press fit.

7.1.5.1 Volume Equivalent Power Consumption

Using a magnet with outside diameter 1.625 in the AMEBA2 unit cell consumes a maximum of 15.2 W 100 Hz and 93.0 W at 1 kHz.

Using the proposed magnet size, a final verification is to compare the power efficiency of the proposed AMEBA2 transmitter to that of an equivalent electrical coil, which is presented in Fig. 7.10. At 1 kHz, the AMEBA2 consumes a maximum of 92.9 W. Compared to the minimum 2161 W consumed by the electrical coil, the AMEBA2's maximum power consumption is 23.4× lower. At 100 Hz, the AMEBA2 uses a maximum of 15.2 W, 142.1× less power. Power consumption of the equivalent electrical coil at 100 Hz was simulated separately, and is not shown in Fig. 7.10.

The plot of Fig. 7.10 is produced using the same approach as for the AMEBA prototype, which is shown in Fig. 5.22.

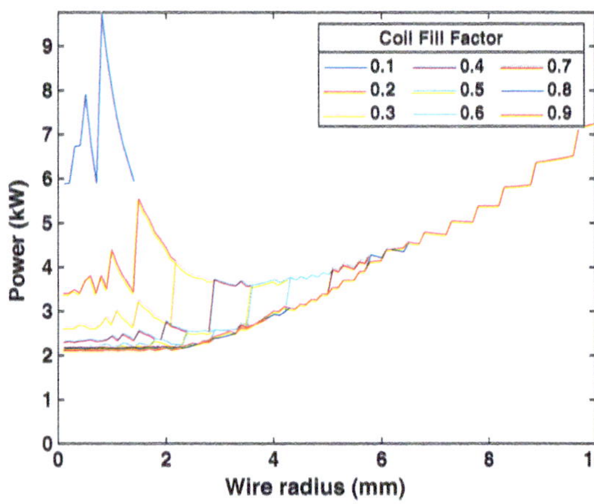

Fig. 7.10 Simulated power consumption (at 1 kHz) of an inductive coil with the same outer dimensions and magnetic moment as the proposed transmitter. Plots are shown for various values of fill factor ρ, where $1 - \rho$ is the ratio of inner to outer diameter for the coil

7.1.6 Critical Speed of the Rotating Assembly

Mechanical systems, like electrical systems, experience the effects of resonance. Applying a voltage off-resonance to a series-tuned RLC resonator causes a current to flow through the circuit in response to the applied voltage. If instead a voltage is applied at the resonant frequency, the series reactances of the circuit cancel and the circuit is incapable of resisting an increase in the energy of the oscillation. Much like this electric circuit, the rotating portion of the AMEBA2 transmitter is subjected to the effects of resonance at a natural frequency of oscillation which causes the elastic forces in the shaft to be completely neutralized [2, p. 496]. Under this condition, the shaft is no longer able to resist a deflecting force and any small disturbance in the axial deflection of the shaft is amplified, causing a vibration.

In the mechanical transmitter, the shaft is suspended between two bearings and the force applied by the magnet on the shaft causes a negligible elastic deformation, known as static deflection (A "DC displacement"). When the shaft is rotated at the same speed as its natural frequency, this deflection is amplified through the effect of resonance. The resonant frequency of a rotating shaft is called the critical speed N_c and is given by

$$N_c = 187.7\sqrt{\frac{\sum W_i y_i}{\sum W_i y_i^2}} \tag{7.1}$$

where W_i are the loads placed on the shaft and y_i are their corresponding static deflections. $W = M \cdot g$ where M is the mass of the load and g is the acceleration of gravity.[1] In the case of a single load, Eq. 7.1 reduces to the rather graceful relation

$$N_c = 187.7/\sqrt{y} \tag{7.2}$$

where y is the static deflection of the shaft in response to the force applied by the load. In the simplest case where the load is modeled as a point source, the static deflection y of the shaft is given by

$$\text{Shaft Supported: } y = \frac{Wa^2b^2}{3EIl} \tag{7.3}$$

$$\text{Shaft Fixed: } y = \frac{Wa^3b^3}{3EIl^3} \tag{7.4}$$

where W is the load applied to the shaft, a and b are the distances between each bearing and the load, E is Young's modulus for the material of the shaft, I is the second moment of area normal to the circular cross section of the shaft[2] ($\frac{\pi}{4}r^4$), and l is the length of the shaft. The shaft is considered supported if very short or self-aligning bearings are used [2], which

[1] We must be careful here to absorb the meaning of the equation in the midst of potentially unfamiliar notation as many symbols have entirely different meanings in the realm of mechanics.

[2] I is also specifically *not* moment of inertia in the common use in relation to mass.

do not resist the deflection of the shaft. The shaft is considered fixed if it is held by long, rigid bearings. The AMEBA2 is supported by long and rigid bearings, however, both cases will be considered for the purpose of complete analysis. Substituting Eq. 7.3 into Eq. 7.2 and using $E = 29 \times 10^6$ psi for a steel shaft gives

$$\text{Shaft Supported: } N_c = 387,500\frac{d^2}{ab}\sqrt{\frac{l}{W}} \tag{7.5}$$

$$\text{Shaft Fixed: } N_c = 387,500\frac{d^2 l}{ab}\sqrt{\frac{l}{Wab}} \tag{7.6}$$

which results in a critical speed of 99.5 rps in the supported case and 199 rps in the fixed case using $l = 3.25$ in for the proposed CAD design and W as the mass of the load and $\frac{2}{3}$ the mass of the shaft as suggested in [2]. Note that as a result of static deflection, the system will tend to vibrate at any speed of rotation, the critical speed being only the point at which such vibration is amplified by resonance. This effect only occurs at the exact frequency of resonance. Approaching the critical speed from either direction (in frequency domain), the shaft begins to lose its resistance to elastic deformation, and vibration increases until it reaches a peak magnitude at the critical speed. Operation at any frequency sufficiently far from the critical speed restores the elastic resistance of the shaft and the vibration [2]. Thus, the rotating shaft (much like an electric circuit) has a quality factor.

In practice, the critical speed is not an upper or lower limit, but rather a region or bandwidth which must be avoided. For operation at frequencies above the critical speed, one approach is to pass through the critical speed by rapid acceleration. Of course, the effect of resonant oscillations at the critical speed could be eliminated to the point where it is not noticeable if the entire assembly were carefully balanced. Most portions of the assembly are carefully balanced; however, balancing the magnet in particular is a considerable challenge (see Sect. 7.3) despite the magnet being the greatest source of unbalance due to poor manufacturing tolerances and varying density which result from the sintering process.

Returning to the subject at hand, this approach is only accurate under the assumption that the load causes a deflection of magnitude y owed to the force applied at a point along the shaft. In practice, shrinking the load onto the shaft would increase its stiffness and reduce deflection, since the load is distributed along the shaft (also reducing deflection), and gyroscopic action of the rotating load tends to straighten the shaft and reduce deflection in motion [2]. Therefore, the value computed may be lower than expected.

As the magnet is very long and thin, stretching 3 in of the 3.25 in span of the shaft, the load is well distributed along the shaft and may likely have infinitely many critical speeds with the lowest being far higher than what has been calculated. The effect of the distributed load may also be quite similar to forming a composite shaft, as nearly the entire portion of the shaft is made of a cross section containing the shaft collar and magnet around the shaft. If the shaft collar is mated to the shaft with an extremely precise fit, it may effectively become a portion of the shaft. In this case, EI (also called flexural rigidity) in Eq. 7.3 may be modeled as $E_s I_s + E_c I_c$ for the composite, where $E_s I_s$ is for the shaft and $E_c I_c$ is for the shaft collar. Using the proposed aluminum shaft coupling, $E_c = 10^6$ psi and the critical speed of the shaft is 521.3 rps. If the experimentally determined critical speed is below this value, heat-shrinking the shaft collar onto the shaft may be an important modification to consider during the development of the realized experimental setup.

Given the information above, and the more terse explanation $N_c \propto \sqrt{EI}$, the critical speed could be further increased by increasing the stiffness of the shaft coupling. Aluminum is specified due to its low weight which contributes negligibly to τ_{req}, and being non-magnetic it is easy to install in the magnet. Other materials may offer greater stiffness if vibration at the critical speed is deemed to be a concern. One such example of an alternative non-magnetic material is Nickel 625 (which some may know as Inconel® 625). Ni625 has Young's modulus 30.1×10^6, and despite having $\sim 3\times$ density it results in an estimated critical speed of 838.4 rps. In this case, the increase to τ_{req} would also reduce the maximum operating speed of the assembly. Other (more expensive) materials that combine high stiffness with low weight include titanium and beryllium.

The final estimator is to consider the case where the rotating shaft is a composite of the shaft, aluminum collar, and the magnet itself. This case is likely inaccurate given neodymium cannot be heat-shrunk and is susceptible to shattering under stress; however, it is likely possible to forcibly hammer an aluminum plug of finely dimensioned outside diameter into a neodymium alloy magnet. Treating the entire rotating system as a composite gives a critical speed of 7572 rps.

7.2 A Three-Phase Motor Driver

7.2.1 Drive Type and Requirements

Inside the DC permanent magnet motor, the stator consists of three phases with windings W_1, W_2, and W_3, respectively. A rotating magnetomotive force (mmf) is produced by driving each phase in a sequence based on the physical winding configuration. This process is known as electronic commutation, since the motor requires an external circuit to perform commutation rather than internal contacts (such as brushes). Six discrete orientations of the rotating field are possible using three-phase windings and a drive source which supplies current in either direction. The rotor can have any even integer number of poles N and is mounted either inside or outside the stator based on the manufacturer's design interests. Internal rotors are preferable when high speed and high dynamic response are required, as the rotor's moment of inertia is minimized. An external rotor produces force farther from the center of rotation (thus producing greater torque) though the increased moment of inertia reduces both speed and dynamic response [3]. This is a subtle optimization left to the motor designer, as the impact varies with the size and scale of the motor.

Here, we focus on configurations having a three-phase stator and an internal rotor with 2 or 4 poles, a depiction of which may be found in Fig. 7.11. The figure depicts slotted windings, though some EC motors use slotless windings instead. In either case, the stator is

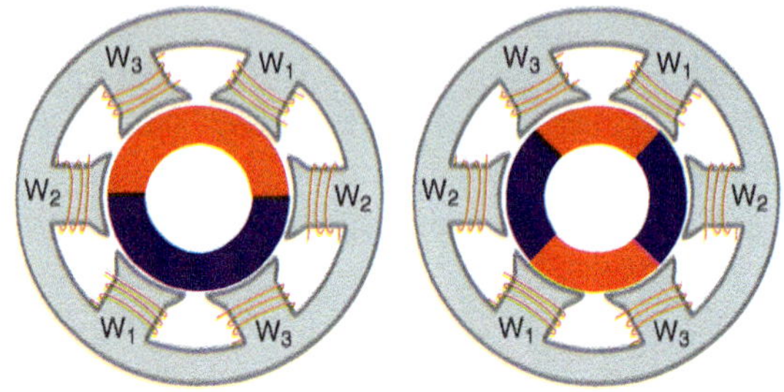

Fig. 7.11 Example rotor and stator configurations with three-phase slotted stator windings (each wound on two opposing teeth) and an internal rotor with 2 (at left) and 4 (at right) poles

internally wired to present as a three-terminal device in a delta (Δ) or wye (Y) configuration (similar to those used in AC induction motors). The DC motor power amplifier must be able to independently drive each terminal high or low to produce the six mmf vectors required for commutation.

Three half-bridge (push-pull) output stages are used for the final output stage, as only one winding of the motor must be energized at a time. There are several critical considerations for a bridge driver. When an H-bridge is used to drive two terminal loads, shoot-through between high and low side drivers during state transitions is a problem and an optimization challenge. Fortunately, the three-terminal load of the stator does not require any one terminal to switch between low-impedance states in a single transition. As an illustration, Fig. 7.12 displays the state transitions between current flows during the commutation sequence. In the first configuration (top-left), current is conducting between terminals A and B, i.e., the power amplifier is driving terminal A high and terminal B low.

To rotate the magnetic field to the next coil in the sequence, terminal C must be driven high (to produce current 2) and terminal B must switch to the high-impedance state. Terminal C began in a high-impedance state and there is no possibility of shoot-through in its driver, and B is simply switching off. De-energizing the windings A⇒B produces a positive voltage spike on B resulting from the sudden interruption of current, which conducts backwards through the high side driver of B, commonly called *freewheeling*.

Three Hall-effect sensors in the motor supply feedback on the current position of the rotor with respect to the stator windings. Referring to Fig. 7.13, unique combinations of A, B, and C correspond to the six required output configurations of the power amplifier. The pattern repeats once per revolution of the rotor for a two-pole motor. The same approach is effective for an N-pole motor having the same winding configuration by repeating the sequence $N/2$ times per rotor revolution. The EC sequence more closely approximates a continually rotating field as N increases, thus increasing the smoothness of acceleration response (and perhaps rate of rotation). In either case, this repeated pattern is referred to as block commutation, which is only an approximation of the ideal continually rotating mmf acting on the rotor. Fortunately, the rotor has inertia, which acts as a low-pass filter and smooths its response.

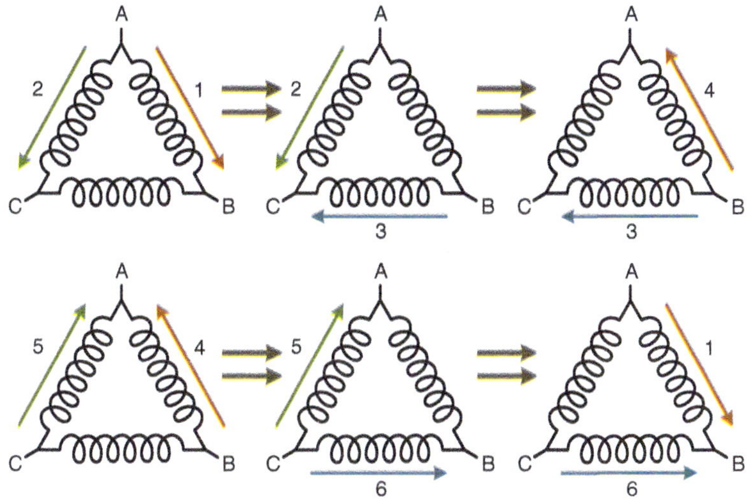

Fig. 7.12 Phase rotation pattern in delta windings of a three-phase EC motor. Current vectors are numbered in order of the commutation sequence; each subfigure depicts a transition to the next state in the sequence

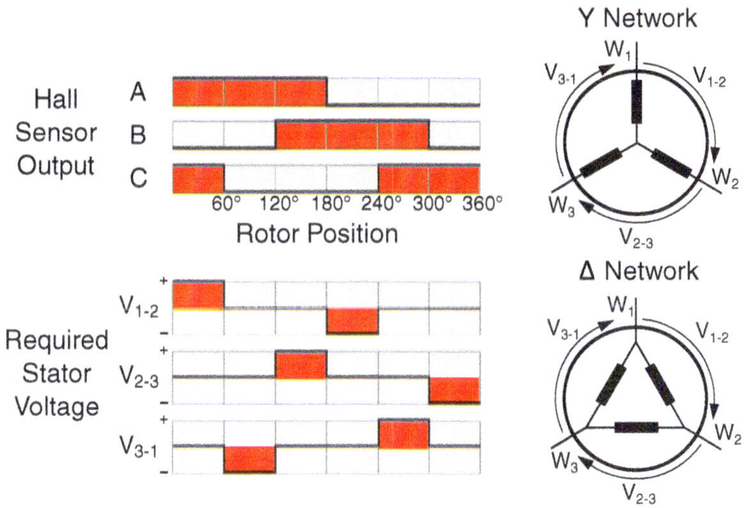

Fig. 7.13 Position feedback and drive voltage patterns for a 3Φ EC BLDC motor with two pole rotor. For an N-pole rotor, the sequence repeats $\frac{N}{2}\times$ per revolution, and indicated rotor positions are reduced by an equivalent factor

7.2.2 Motor Controller Board

An AMEBA2 **Mo**tor **C**ontroller (AMOC) was designed to drive the three-phase electronically commutated (EC) windings of the maxon ECX motor. The NMOS half bridge uses International Rectifier IRLML6344 power MOSFETs with low $R_{DS,on} < 29\ \text{m}\Omega$ and $Q_g = 6.8$ nC to minimize both conduction and switching power loss. The FETs must source/sink up to ~4.2 A (see specifications given in Sect. 7.1.1) continuous current and withstand freewheeling transients.

The controller implements a Texas Instruments DRV8305 three-phase gate driver. The IC includes charge pump-based floating gate drivers for the high side of each half bridge and amplifiers for current sense resistors on each phase. A host of fault protection and diagnostic features are also implemented on chip including configurable maximum gate drive current, armature current, and dead time between phases which are of particular interest. These features allow the control system to drive the motor by varying only the frequency and duty cycle of three PWM signals and their complements which are supplied to the IC. All other parameters are configurable by SPI. Fault diagnostic information is also provided over SPI which simplifies debugging in the field.

A 6-PWM control scheme is used to control the commutation sequence for the greatest degree of control over the motor driver/power amplifier. In this configuration, 3 pairs of 2 independent high side (IN_H) and low side (IN_L) inputs are used to independently configure the state of each half bridge as high, low, or high-Z as required. All signals are active-high, abstracting the digital controller from the high side floating gate drive topology. In addition, the DRV8305 in the power amplifier drives both gates low when IN_H and IN_L are both high for any given bridge, ensuring robustness against devastating shoot-through effects in the event of an unexpected transient.

For completeness, observe from Fig. 7.14 the timing diagram for a single transition of one transistor. A transition from the controller corresponds to the charging (or discharging) of a single transistor gate through a current source/sink, during which time the current sink for the other transistor in the bridge is switched on to absorb any transient charge which appears at the gate, effectively preventing spikes from causing shoot-through in the bridge which may damage the power FETs. This is a feature of the IC which cannot be externally disabled and, while useful, also introduces the potential for a catastrophic failure of the single chip driver. If a large enough transient is discharged through the current sink, the instantaneous power dissipation has the potential to destroy the charge pump circuit. This condition has been observed in practice when a motor driving a heavy load requires a large current to be supplied to overcome the force of static friction. When the system begins to rotate, the sudden reduction of drag results in a rapid acceleration of the rotating body and a rise in back emf which may trigger the DRV8305 IC to absorb transients large enough to damage or destroy the chip.

Interfacing between the motor driver board and the control system uses a 26-pin header. PWM signals and IC control signals are supplied to the board by the control system and

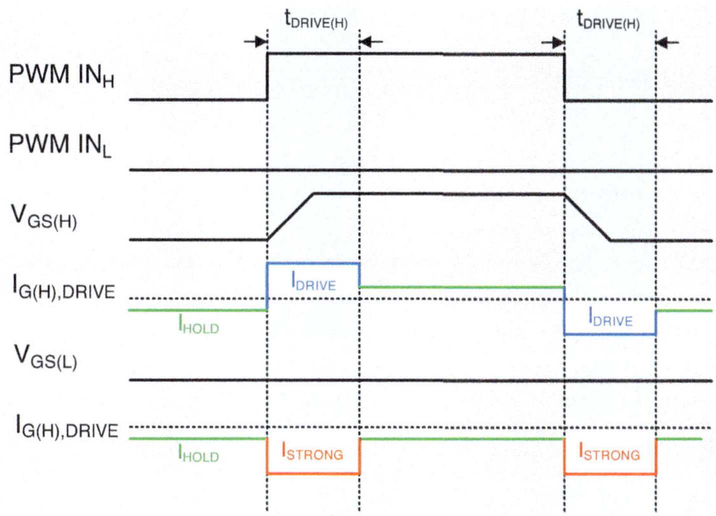

Fig. 7.14 Half-bridge gate drive scheme showing switch-on and switch-off events

Fig. 7.15 Assembled AMOC v1.0 on the lab bench

outputs from all three encoders are buffered inputs to the control system. A two-way SPI interface is used to communicate with the PCB.

The assembled motor controller is shown in Fig. 7.15. The PCB is fabricated using 2 oz copper for improved power efficiency and heat dissipation; this also decreases the footprint required for high power traces of equivalent resistance. Power to the motor is supplied directly to the board from a 24 V source, and communication with the control system is accomplished over a single-pin header.

7.2.3 Revised Motor Controller Board

On the workbench, the first version of the power amplifier held up well for testing unloaded motors and developing some of the software in Chap. 8, though when the motor is loaded the controller exhibited intermittent failure of the driver IC. Real-world testing suggests that the controller typically handles transient events without issue, though one or more gate drive FETs are observed to fail together with the IC. It is thought that either the FETs fail due to generator-mode voltage spikes while in the high-impedance state (shorting the gate current source to V_{DD} or V_{SS}) or spurious currents induced by the impulse response of the heavily loaded motor are improperly handled by the active freewheeling technique of the drive. In either case, better performance can be found at the state of the art in switching technology, and the board was revised to use cascode structure GaN FETs (TPH3212PS, Transphorm USA), each of which consists of a gallium nitride high-electron-mobility transistor (GaN HEMT) and silicon MOSFET.

The revised board (AMOC v1.1) is shown in Fig. 7.16 and also includes a transient snubber and improved heatsinking through large package FETs. Transient spikes are suppressed using ferrite beads specified by the GaN FET manufacturer, though we note that this is against the advice of the TI IC datasheet. The $R_{DS,on}$ is increased to 85 mΩ and

Fig. 7.16 Assembled AMOC v1.1 with improved thermals and transient tolerance

$Q_g = 14$nC compared with the previous iteration of the controller though the output stage is now rated to $|V_{GS}| < 18$ V and $V_{DS} < 650$ V.

7.3 Machining and Assembly

The motor mount, bushing mounts, and base component were CNC milled from acetal homopolymer, specifically DuPont Delrin 150. The CNC service holds a tolerance of ±0.005″; critical features such as dowel pin and bolt holes were left undersized by 0.010″. Components are shown in Fig. 7.17 with the motor installed to verify fit. This represents the first physical component of the AMEBA2 transmitter.

Guide holes in bushing and motor mounts were drilled out to a clearance fit for #10–24 machine screws. Alignment holes were reamed to a press fit with nylon dowel pins. The vibration of the driven shaft (due to imbalance or resonance) will be transmitted through the assembly, so bushing mount block holes were reamed for a tight press fit using O-rings. Here, the O-rings serve as the mechanical equivalent to a damping resistor in an RLC resonator. The O-rings also reduce the force applied to the shaft in response to small deflection angles. Air bushings were installed, the shaft was cut down to size, and axial alignment was checked again before tightening the retaining screws. The shaft was then coupled to the motor. Figure 7.18 shows a photograph of the assembly at this stage.

Producing the coupling between the magnet and shaft is an interesting challenge since standard high-precision measurement tools (micrometers, dial indicator) are sensitive to strong magnetic fields. A piece of circular aluminum stock was turned down to 0.010″ over the ID of the magnet and the fitment was checked by hand. In successive passes, 0.0005″ was removed on the OD of the coupling and fitment was checked again. This process was repeated until the aluminum coupling can be gently pressed into the magnet.

Fig. 7.17 Rapid turn CNC components of the assembly as received from the CNC service

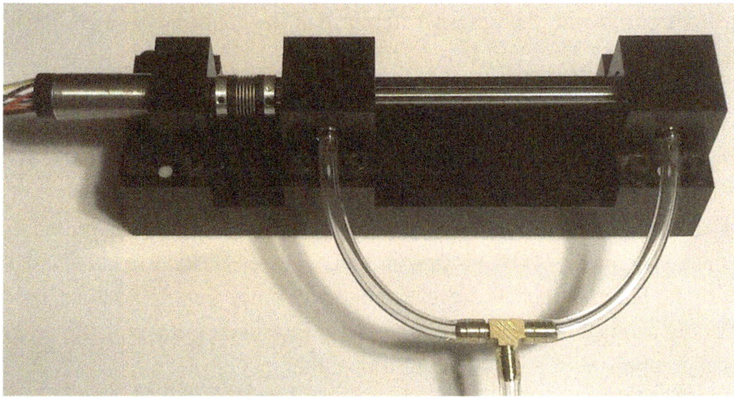

Fig. 7.18 After stage 2 machining, all components except for the magnet and its shaft collar (coupling) have been fitted

7.3.1 Alternative Methods for Centering the Magnet

Ideally, the center hole in the coupling would be drilled by first permanently mating the coupling to the magnet, then drilling a hole which centers either the outside diameter of the magnet or the center of mass along the axis of rotation.

Centering the outside diameter of the magnet minimizes air friction and is a first-order approximation to the center of mass. This would be accomplished by fixturing the magnet from the outside diameter and turning between centers to drill the centered hole. Vibration is of greater concern than air friction (as it increases the power requirement and frequency instability in ways which have not been discussed). The magnet could be centered along the center of mass by identifying the center of mass on the XY plane normal to the axis of rotation, supporting the magnet in a vertical mill, and drilling the center hole this way.

Both methods described are somewhat out of reach as each requires fixturing a strong permanent magnet in a lathe or mill, respectively, both of which are made of cast iron and steel components. Instead, a hole is drilled through the center of the coupling after turning down the OD, before removing it from the lathe. Assuming constant density, the coupling is then to be mounted on the shaft along its center of rotation and center of mass. However, the resulting composite of the magnet and collar are not centered due to limited manufacturer tolerancing of the magnet itself. It would be possible to achieve proper centering of the magnet by first producing a non-conductive jig to support the magnet, then using this jig to hold the magnet in the machine tool.

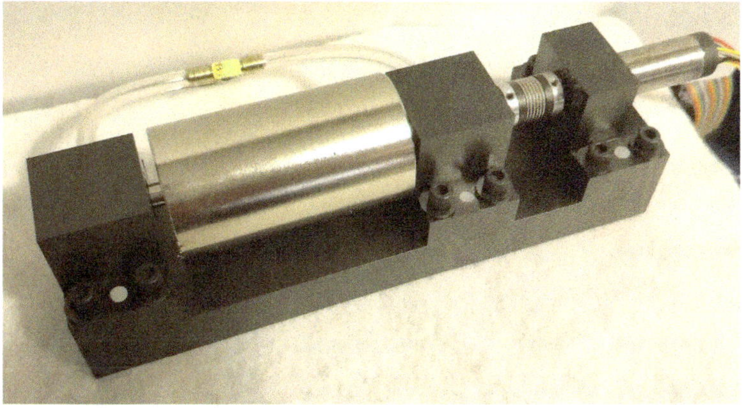

Fig. 7.19 AMEBA2 unit cell assembly completed to initial design specifications

7.3.2 Returning to the Bigger Picture

The completed mechanical assembly is shown in Fig. 7.19. While functional, tests showed that the friction of rotation of the proposed transmitter is prohibitively high. Within the first minute of rotation, the excess power required by the motor resulted in overheating of the windings.

To understand the causes of such high friction, recall that the rotating shaft in AMEBA2 must be fit to an exceptionally high degree of precision. The inside diameter of the air bushings is specified as $\phi\ 0.2507^{+0.0002}_{-0.0000}$ inches as supplied. The manufacturer supplied a shaft with diameter $\phi\ 0.25^{+0.0000}_{-0.0007}$ inches and straightness guaranteed within 0.0003/in. Such tolerances result in a significant design problem. If the two bushings are perfectly (to 0.0000″) aligned on the center axis of rotation, then the worst-case minimum clearance between the shaft and the farthest point on each bushing is −0.0010″. That is, it is possible for a combination of two bushings and one shaft to have an insufficient air gap before accounting for the misalignment of the bushing supports, which will of course not be 0.0000″. The best-case clearance (assuming perfect shaft straightness, minimum shaft diameter, and maximum bushing ID) is 0.0016″.

Replacing the shaft with an ultra precision 303 stainless steel shaft (Nordex, Inc) with diameter $\phi\ 0.24975^{+0.0000}_{-0.0001}$ inches straight to within 0.0001/in improves the situation, giving a clearance range 0.0004″ − 0.0013. Nevertheless, there still remains the issue of absolute alignment.

A heatsink was machined from 6061 aluminum to improve thermal power dissipation in the motor, as shown in Fig. 7.20. However, this effort proved fruitless as the rate of internal heat dissipation to the casing was still too great.

Fig. 7.20 A generously sized heatsink dissipates large amounts of heat from the small motor

7.3.3 "Fit by Feel"

Misalignment of each bushing is a function of two planar displacements and two angular displacements, and there are no convenient reference points. The height of the shaft with respect to the assembly baseplate may vary based on the surface flatness of the baseplate. Indirect measurements (i.e., through multiple components) are ineffective here. Accordingly, the inside diameter of each bushing block was measured using an inside micrometer with accuracy 0.0001 (Mitutoyo) calibrated against a precision reference standard with inside diameter 0.70002″ (INSIZE USA). A one-off alignment tool was made to suit the dimensions of the assembly. A portion of circular 6061 aluminum stock having overall length 8″ was turned to an outside diameter 0.001″ lower than the inside diameter of each bushing block. The resulting tool has a taper < 0.0007 over the distance between outer walls of the support blocks and eccentricity < 0.0002. Dowel pin holes in the base were checked for straightness using the tool. One dowel pin must be removed from each support block to allow the tool to fit smoothly through the assembly, indicating which positions are out of alignment.

An unloaded shaft was hand polished to maximize surface smoothness, then fitted between the air bushings with the motor disconnected. The support block which fixtures each bushing was loosened slightly and both height and angle of each block were corrected by installing hand-trimmed pieces of 0.0005″ shim stock which are layered around the base component as needed. With the air supply on, when the shaft is properly aligned without interference in either support it tends to move along the axis of rotation (due to the weak but nonzero net force of gravity and interaction between air currents escaping the bushing from either side).

The motor was then connected to the driven shaft. When driving the unloaded shaft at a constant velocity, the power consumption of the motor varies linearly with assembly friction. Fine adjustment to each bushing was performed by applying thread locker to all four bolts holding the support blocks to the baseplate, and adjusting each until the minimum power consumption is observed.

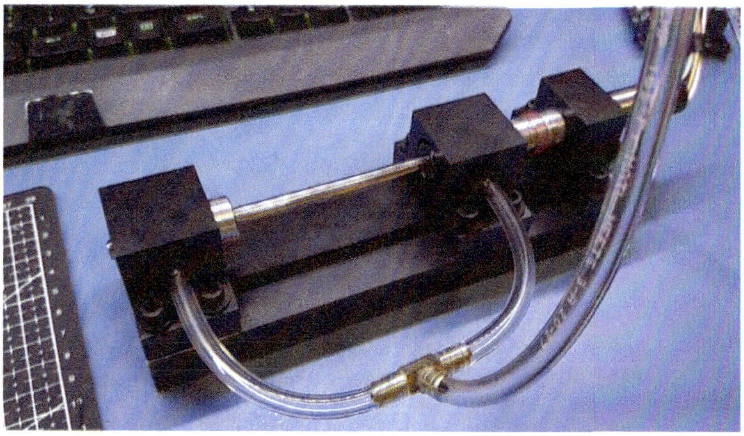

Fig. 7.21 Fit-by-feel assembly using the miniature jaw coupling

Next, the bellows coupling was replaced with a jaw coupling as shown in Fig. 7.21. Bellows couplings are ideal when accurate transmission of torque between shafts with large angular displacement is called for. However, they have very low tolerance for planar displacement (due to the single-piece design), and angular misalignment in the coupling applies a load torque to the motor. On the other hand, the jaw coupling has greater absolute tolerance for planar displacement and reduced load torque in response to angular misalignment. The jaw coupling still contributes some torque, though reduced torsional stiffness in the spider minimizes the effect. Removing the jaw coupling spider minimizes coupling effects and allows the alignment technique to be applied for the load shaft alone; the spider is reinstalled and the motor mount aligned afterward. With the shaft being supported by air cushions, the lathe is only somewhat useful as contact between a dial indicator, and the shaft can interfere with the alignment of the latter. It was used to check the eccentricity of the jaw coupling before and after the installation of the spider.

Due to the loading imposed by the magnet, the shaft was not well supported by the air bushings when not freely rotating. When the magnet begins to rotate from rest, an air film develops in the bushings and the load torque due to static friction in the bushings drops to near zero. The required instantaneous startup torque, in combination with the development of back emf, creates a transient fault condition which motivates the use of a minimally sized magnet. The eccentricity of the rotating mass and remaining misalignment in the rotating system still contribute load torques after balancing. These factors also motivate reducing the maximum operating speed and replacement of the motor with one which supplies greater torque. The final assembly reflecting these changes is shown in Fig. 7.22 with the previously mentioned alignment tool.

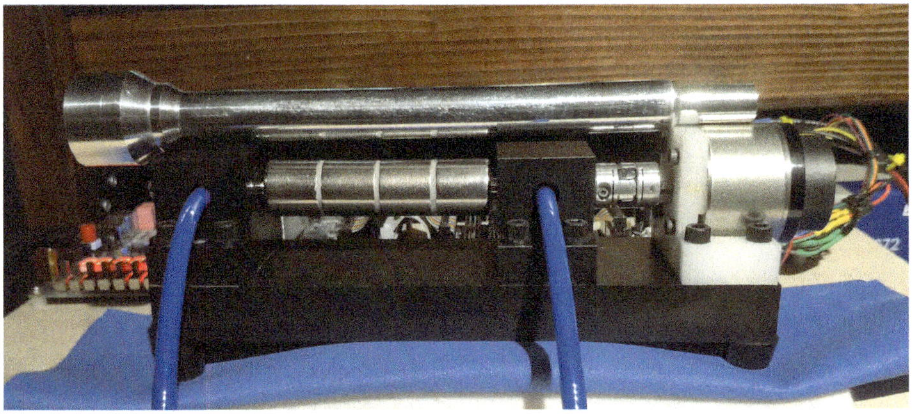

Fig. 7.22 Final unit cell using a minimally sized magnet, shown next to the shaft alignment tool

References

1. Galton, I., & Weltin-Wu, C. (2018). Understanding phase error and jitter: Definitions, implications, simulations, and measurement. *IEEE Transactions on Circuits and Systems I: Regular Papers, 66*(1), 1–19.
2. Marks, L. S. (1951). *Mechanical engineers' handbook*. Incorporated: McGraw-Hill Book Company.
3. Reichert, T., Nussbaumer, T., & Kolar, J. W. (2009). Torque scaling laws for interior and exterior rotor permanent magnet machines. In *The IEEE International Magnetics Conference (INTERMAG 2009)* (Vol. 3, No. 1).

8 Transmitter Control System

8.1 System Overview

The transmitter control system is organized into four main components:

- User-facing console;
- Digital controller;
- Power amplifier;
- AMEBA2 unit cell.

The first two components are introduced in this chapter and fully integrated with the power amplifier and mechanical unit cell. In the classical controller model, the user-facing console is the interface between the human operator and controller. The digital controller is an application-specific digital logic circuit synthesized on a field programmable gate array (FPGA). The power amplifier and AMEBA2 unit cell are the electrical and mechanical components of the plant. For the most part, the impact of the power amplifier on the control system is minimal and we find that the plant may be approximated as the unit cell alone.

Control of the AMEBA2 transmitter is accomplished by means of a command-line interface (CLI) written in C. The CLI runs on the hard processor system (HPS) of an Intel (formerly Altera) Cyclone V system-on-chip (SoC), which offers a unique combination of a high-performance processor (a dual-core HPS) and programmable logic (an FPGA) in the same chip. The HPS runs a full-fledged GNU/Linux operating system (Debian Stretch) and is capable of interfacing with the FPGA fabric through memory-mapped parallel IO (PIO) and direct memory access (DMA) means.

The somewhat simple combination unleashes great power in the form of a direct interface between computer software and hardware-level digital logic. A C console is convenient to generate communication-ready signals from arbitrary input and provide a high-level inter-

J. S. Glickstein and S. Mandal, *Electromechanical Transmitters for ELF/VLF Radio*, Synthesis Lectures on Engineering, Science, and Technology,
https://doi.org/10.1007/978-3-031-15123-1_8

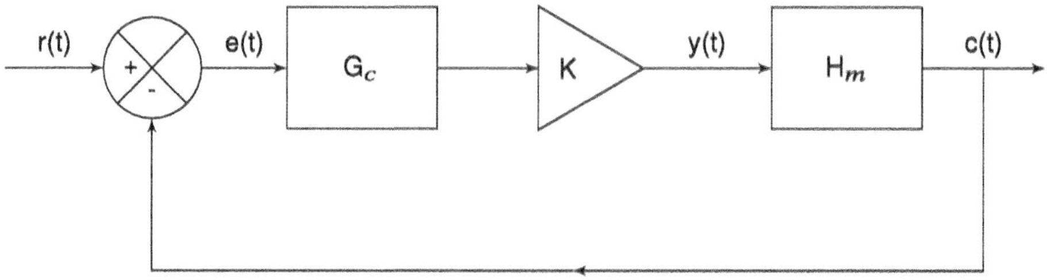

Fig. 8.1 Established control theory notation describing input-output characteristics of the system

face to the transmitter. Meanwhile, control of the plant is implemented in the FPGA fabric and yet remains accessible for software-based adjustment of parameters. While the FPGA's exacting control over timekeeping is a major advantage, the implementation of user interfaces and control of memory is limited. Augmentation with high-level software through the HPS is used to realize an effective route to this end.

It is here during the development of the control system where the realm of theory first meets the realm of reality. Some design adjustments are noted throughout, elucidated by the performance of the system test bench as compared with the idealized models which have been used up to this point.

Figure 8.1 displays the principal notation used to describe generally the components of the system formed by closed-loop controller and plant. Input $r(t)$ is a target speed—in units of revolutions per second, or Hertz—fed from the modulator on the software side using the embedded system. The controller and motor have transfer functions G_c and H_m, respectively. The controller output $y(t)$ and measured parameter $c(t)$ of the plant are in the same units as the input.

8.1.1 Hardware Test Bench

The Cyclone V is implemented using a Terasic DE1-SoC development board which provides an abundance of peripheral interfaces for easy integration into the experimental setup. Two 40-pin headers provide fan-out for 72 IO pins reserved for general-purpose use, enough to interface with up to four EC motor drivers and control up to four mechanical transmitters simultaneously. A hardware test bench including one mechanical transmitter is shown in Fig. 8.2, with an interface for additional parallel units maintained for future use.

The HPS uses an off-chip 1 GB DDR3 SDRAM for program memory and shares its memory controller with the FPGA fabric for memory-mapped data transfer (direct memory access, DMA). The two are also linked through an AXI bus (Advanced eXtensible Interface), and parallel IO (PIO) ports can be used for lightweight communication links between components.

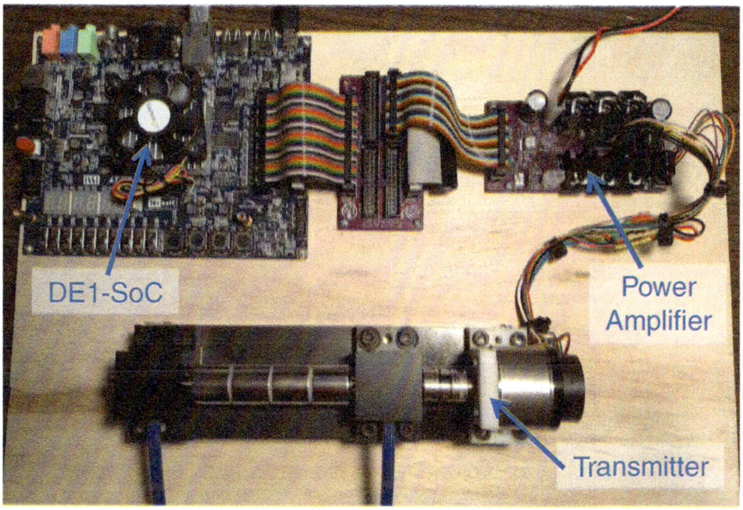

Fig. 8.2 Hardware test bench for design and debugging purposes. Power and air supplies not shown

User-level communication is enabled through a Bash shell accessible through a universal asynchronous receiver/transmitter (UART) or secure shell (SSH) interface, using an off-chip USB-UART and Ethernet controller, respectively. While the GNU/Linux operating system on the HPS can run a full desktop environment such as LXDE, it is largely unnecessary.[1] Furthermore, a graphical interface does not only consume HPS resources—the SoC does not have a dedicated video controller and one must be synthesized on the FPGA (consuming resources) if needed. It is best to set up the design with the intent to minimize resource consumption to reduce the possibility of running out of resources during the development process, and save headroom where possible for when it is needed most.

An equivalent block diagram of the control system layout is supplied in Fig. 8.3 to highlight key components.

The digital controller includes submodules for debugging/self-test features in addition to live operation mode. In debugging mode, controller parameters may be changed on-the-fly and process variables are fed back to the console—all experimental data reported in this chapter has been recorded by the controller itself. All inputs to the power amplifier are treated as manipulated variables and the controller is used to run experiments for experimental characterization of the plant. On the other hand, a live mode is a demonstration of operation in real-world scenarios such as a search and rescue mission, where the console presents a simple text chat interface.

[1] From the point of view of typical GNU/Linux users.

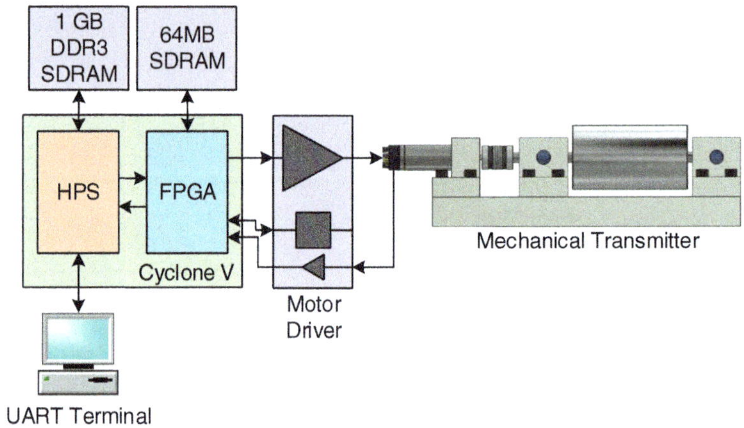

Fig. 8.3 Transmitter control system consisting of a hard processor, FPGA, random access memory (RAM), and the hardware-level motor driver of Sect. 7.2.3

The DE1-SoC board supplies a VGA digital-to-analog converter (DAC) and 64 MB SDRAM (intended for storage of a VGA framebuffer) routed directly to FPGA GPIO pins, the latter of which is convenient to repurpose as a low-latency cache for real-time data collection.

8.1.2 A Model for the Plant

Before diving into the development of the control system, let us model the mechanical transmitter's input-output characteristics. First consider the motor—the input to the motor is the voltage $v_{in}(t)$ from the power amplifier, which produces a current $i(t)$ in the motor windings:

$$v_{in}(t) - v_{emf}(t) = L\frac{di(t)}{dt} + Ri(t) \tag{8.1}$$

where $v_{emf}(t)$ is the back emf in the armature. During operation under load, the motor produces a torque $\tau(t) = ki(t)$. For the complete transmitter, $\tau(t)$ is given by Eq. 6.19. Writing Eq. 6.21 in more general terms,

$$\tau(t) = J\frac{d\omega(t)}{dt} + D\omega(t) = ki(t) \tag{8.2}$$

where D is a damping coefficient owed predominately to air friction. Taking the Laplace transform of Eqs. 8.1 and 8.2 (substituting $v_{emf}(t) = k\omega(t)$ in Eq. 8.1) gives

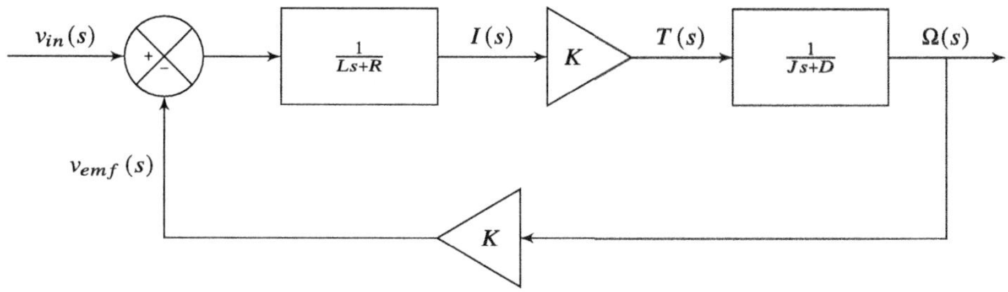

Fig. 8.4 Block diagram representation of the input-output characteristics of the motor

$$T(s) = (Js + D)\Omega(s) = KI(s) \tag{8.3}$$

$$v_{in}(s) - K\Omega(s) = (Ls + R)I(s). \tag{8.4}$$

Eliminating $I(s)$ from both equations gives

$$\Omega(s) = \frac{K}{(Js + D)(Ls + R)}(v_{in}(s) - K\Omega(s)). \tag{8.5}$$

An equivalent block diagram of Eq. 8.5 is produced in Fig. 8.4 above. The feed-forward path is intentionally separated into three fundamental components for clarity—two poles and a constant gain (all of which are familiar to readers of Chap. 6). The input $v_{in}(s)$ is low-pass filtered to produce a current $I(s)$. The current produces a torque $T(s)$ which is then low-pass filtered to yield the current angular velocity $\Omega(s)$ of the plant. Induced e.m.f. in the rotor is essentially negative feedback from output to input. To find the transfer function $\frac{\Omega(s)}{v_{in}(s)}$, Eq. 8.5 is simplified using Black's formula:

$$\frac{\Omega(s)}{v_{in}(s)} = \frac{\frac{K}{(Js+D)(Ls+R)}}{1 + K\frac{K}{(Js+D)(Ls+R)}} = \frac{K}{(Js + D)(Ls + R) + K^2}. \tag{8.6}$$

8.1.2.1 Simulation

A Simulink test bench for the motor model of Eqs. 8.5 and 8.6 is shown in Fig. 8.5. For now, the controller is simply a pre-defined signal vector with values corresponding to the full-scale output of the FPGA controller. Correspondingly, the power amplifier block is a gain block which produces a value in the range 0–24 representing the voltage produced across the motor terminals.

Using varied values for J (considering a selection of permanent magnets as described in Chap. 6), the step response of the model motor is observed on the Simulink scope to

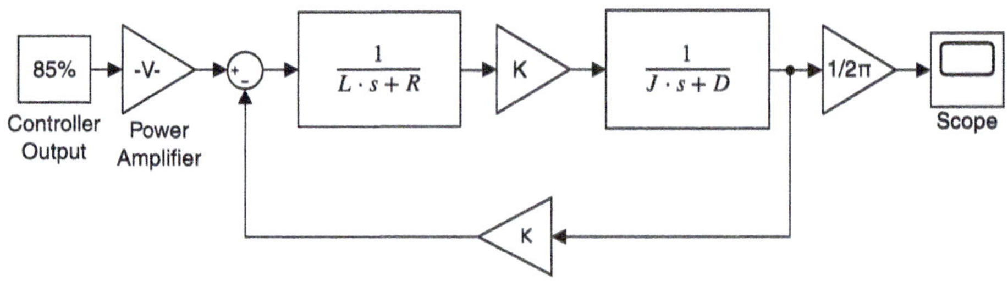

Fig. 8.5 Simulink model containing basic input-output characteristic

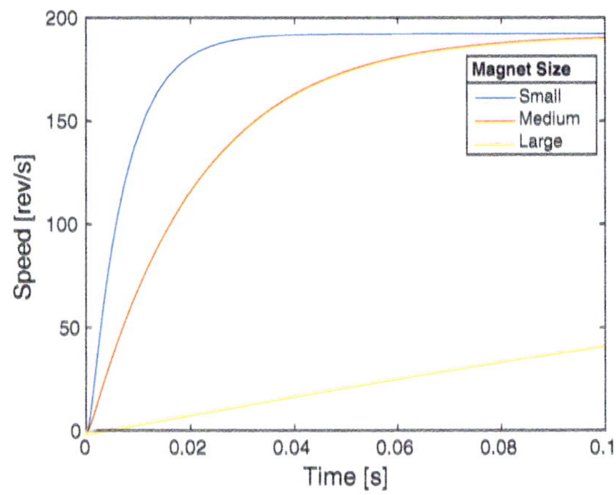

Fig. 8.6 Simulated response of the motor (at rest) to a step input under various load conditions

demonstrate functionality. The results are summarized in Fig. 8.6. The maximally sized magnet's response is on a scale $\sim 25\times$ greater than what is shown, which has been suppressed to show details in the more relevant curves.

8.1.3 Remarks

From Eq. 8.6, it may be observed the motor is most purely a second-order system with electrical and mechanical time constants L/R and J/D, respectively. If (for the sake of approximation) the effect of drag is taken to be negligible ($Js \gg D$) and the inductance of the motor's windings is further taken to be insignificant ($Ls \ll R$), the plant may be approximated as a first-order system:

$$\frac{\Omega(s)}{v_{in}(s)} = \frac{K}{RJs + K^2} \tag{8.7}$$

which results in the time constant already reported in Sect. 6.2.4.1 after a thorough review of the present literature [2]. A more realistic model would include drag D as the effect of air friction is known to be nonzero with its value found in Sect. 6.3.3, which still allows the plant to be optimistically modeled as a first-order system:

$$\frac{\Omega(s)}{v_{in}(s)} = \frac{K}{RJs + (RD + K^2)} \tag{8.8}$$

which remains a reasonable approximation of Eq. 8.68.68 when the purely mechanical pole is the dominant pole in the plant. This is generally the case for most motors and most loads, an expectation which holds true for the AMEBA—a rather heavily loaded system.

Driving the motor terminals with a current source instead of a voltage source would eliminate the electrical time constant entirely, which requires taking a different approach to implementing a power amplifier. When the motor is sufficiently loaded and the prior approximations hold, it is generally simpler to use a voltage source and deal with the second-order system. These criteria are recommended considerations for the selection of power amplifiers and motors in future work.

8.2 Digital Controller

8.2.1 Driving the Motor

The motor is controlled by pulse width modulation (PWM), where the on time τ of a rectangular pulse train $p(t)$ with period T and frequency $f_{PWM} = \frac{1}{\tau}$ is varied between zero and T. Accordingly, the duty cycle D of the pulse is $\frac{\tau}{T}$, and an equivalent series which describes the pulse train for constant D is

$$p(t) = A\sum_{k=0}^{\infty} u(t - kT) - u(t - (k + D)T) \tag{8.9}$$

where A is an arbitrary gain constant. The average value of $p(t)$ over a long series of pulses with constant D is then $D \times A$. Between stages of digital logic, the instantaneous value of $p(t)$ remains one of two distinct values $[0, A]$. However, the armature windings of a DC motor form an LC low-pass filter with cutoff frequency

$$f_c = \frac{1}{2\pi\tau_e}.$$

Provided the PWM carrier frequency is sufficiently lower than the cutoff frequency of the windings, the current developed in the windings is a result of only the average or "DC" value of the modulated input voltage. As with other first-order filters, this is generally taken to be the case for $f_{PWM} > 10 \times f_c$.[2] If there was a need for further support, we consider that the output of the power amplifier is filtered again in the mechanical domain by the rotating magnet. We might be left to ponder what happens when the PWM signal is time-variant and the interested reader is directed to Appendix A.

Selection of f_{PWM} is bounded by several parameters. To ensure the high-frequency components are filtered by the motor windings, the frequency should be at least an order of magnitude greater than $\frac{1}{2\pi\tau_e} = 2.35$ kHz. Frequencies up to 20 kHz (occasionally higher) can be audible to the human ear, which also forms a lower limit to avoid disturbance of the operator and nearby individuals.

Upper limits on f_{PWM} are based on the switching rate of the GaNFETs in each half bridge. The TPH3212PS used in the power amplifier has a gate charge of 14.6 nC, requiring 292 ns to charge with the 50 mA minimum current supply from the charge pump IC on the power amplifier. This imposes a minimum PWM duty cycle $D_{min} = \frac{T_{SW}}{T_{PWM}}$. At a switching frequency of 30 kHz (period 33.3 μs), this is 0.9% of each PWM cycle. At 200 kHz (period 5 μs), this increases to 5.8%. Another factor is the average load placed on the charge pump, rated to a maximum of 30 mA_{RMS}. A first-order approximation $Q_G N_{SW} f_{PWM}$ for the charge pump load estimates 2.6 mA_{RMS} and 17.6 mA_{RMS} for 30 and 200 kHz, respectively. Generally speaking, it is preferable to use as low a frequency as possible while meeting the required rate of dynamic response as this minimizes switching power loss.

8.2.1.1 Digital PWM Signal Generator

A uniform sampling modulator is generated by a free-running binary up-counter of configurable width N which is clocked by a high-speed reference f_{REF}. The input to the signal generator module is an N-bit binary register which specifies the duty cycle D as a number d in the range 0–2^{N-1}. The output of the module is the result of logical comparison between d and the value n_{ctr} of the binary counter which resolves to 1 for $n_{ctr} < d$ and 0 otherwise. The result is an output signal with duty cycle

$$\frac{d}{2^N}\%$$

and carrier frequency

$$f_{PWM} = \frac{f_{REF}}{2^N}.$$

We face a trade-off between minimum switching speed and maximum PWM resolution due to hardware limitation on the maximum f_{REF}. Our particular Cyclone V is specified for a maximum stable clock frequency 500 MHz at the output of a phase-locked loop (PLL) module before considering the effects of process variation and clock skew. We take the maximum f_{REF} to be 450 MHz to avoid potential issues arising from operation at the

[2] By f_c, here we mean the cutoff frequency of the subsequent low-pass filter (in this case the motor windings).

Table 8.1 Switching frequency and minimum duty cycle resolution for several selections of N at $f_{REF} = 450$ MHz

N	f_{PWM}	δD
16	6.87	0.002
14	27.5	0.006
12	109 863	0.024

maximum bound of the datasheet. In layout, the reference PLL is placed as close as possible to the PWM signal module to keep the clock stable and satisfy timing constraints.

Values at several points are given in Table 8.1. Assuming linear response, the step resolution in terms of transmit frequency is $k_e \delta D$ where δD is the step size for D. For $N = 14$, the value is below 0.066 Hz for both maxon ECX16 and EC-i motors when unloaded.

8.2.2 Feedback and Speed Measurement

A top-level block diagram of the hardware design is shown in Fig. 8.7. Two methods to monitor motor speed have been implemented within the FPGA—a forward counter and a reciprocal counter. Both approaches utilize the quadrature and zero position signals supplied by an encoder. The encoder supplies outputs as 5 V push-pull outputs to the power amplifier board. There they are routed through transient voltage suppression and high-speed 5.0–3.3 V level shifters before reaching the FPGA.

At the input to the FPGA, data from the encoder must cross from the motor's asynchronous domain into the global clock domain (100 MHz) of the controller. This is easily accomplished by D flip-flops, given the maximum frequency 1.024 MHz of a 1024 ct/rev encoder in the ECX16L motor rotating at 1000 rev/s. All that remains is to debounce and deglitch the inputs to separate valid transitions originating from the rotating encoder wheel passing through the feedback sensor from bounce conditions caused by imperfections in the encoder, and transient spikes caused by voltage noise induced in the wires between the controller and the plant.

The deglitch module implements two sample-and-hold blocks, A and B in Fig. 8.8, which retain the last validated state of each encoder channel. Valid transitions are defined as those which are followed by no activity on either input for a minimum number of clock cycles following the transition. To separate these transitions from transients, all state transitions on either input channel trigger a reset of the free-running counter which then disables the sample-and-hold blocks and freezes their outputs. Data is held until the next positive edge of the global clock when the enable signal is also asserted; the enable signal is asserted when the input from the plant has remained stable for the specified *minimum state* and the free-running counter has been allowed to run without being cleared.

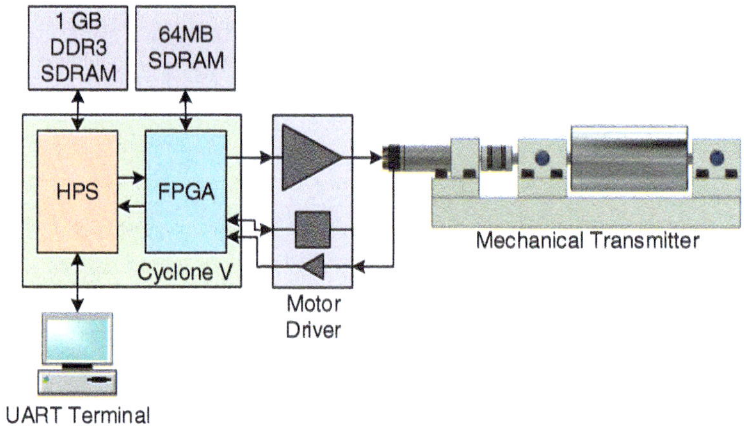

Fig. 8.7 Structure of the top-level hardware design showing core modules (light blue) and connections to other system components (purple). Thin black lines represent specific signals (width not specified) between components. Thick blue arrows convey general interface organization

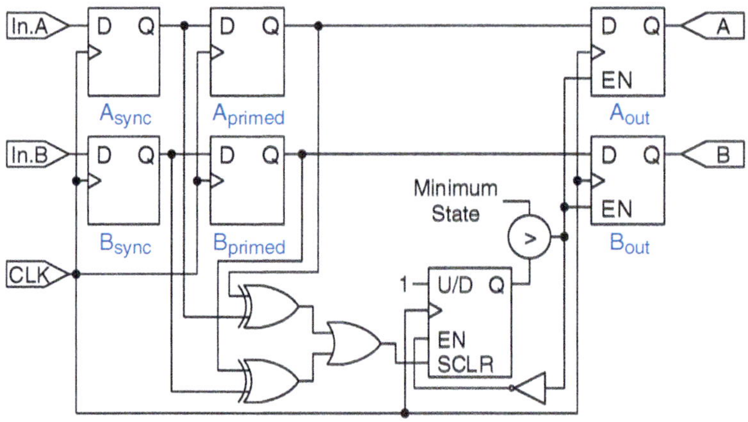

Fig. 8.8 Synchronous logic which samples external encoder feedback and performs validation of asynchronous transitions

At the defined maximum speed of the plant (1000 rev/s) using a quadrature encoder with 1024 ct/rev, the shortest possible time between transitions on either channel is 488 ns or 49 clock ticks on the 100 MHz clock. At 512 ct/rev, the minimum state is doubled, resulting in greater noise immunity without sacrificing the speed of operation. In practice, the transient suppression technique implemented on the controller is shown to be highly effective using either clock, though it is possible to generate a clock with a period as low as 2 ns on the FPGA should the need for noise rejection with greater time resolution arise.

Alternatively, the same effect may also be achieved by reducing the maximum speed of the plant by a factor of $\frac{1}{2}$, down to 500 rev/s. These adjustments are worth consideration in future work should a particularly noisy power supply or long wires between the motor and controller be encountered. Either condition may cause too many transients to be registered by the controller. While replacing the encoder in favor of one which provides less feedback is not feasible during an emergency situation in the field, a transmitter with a large enough dynamic range would permit reducing the maximum transmit frequency to a level which allows for continued operation.

8.2.2.1 Forward Counter Approach

A forward counter simply counts the number of encoder ticks N_{meas} observed in a fixed time interval T_s, supplied by a sample clock f_s. The forward counter then reports the measured speed $n[k]$ as

$$n[k] = \frac{N_{meas} f_s}{N_{enc}} \tag{8.10}$$

where N_{enc} is the number of ticks per full revolution of the motor. Multiplication and division are potential challenges for the FPGA; fortunately, the selection of a power-of-two sample rate and the use of an encoder with a power-of-two number of ticks make this a simple bit shift operation. The only complication is then how to deal with fractional outputs, easily accomplished by storing the results in fixed-width binary format.

Ease of implementation makes the forward counter approach to speed measurement attractive, especially when the FPGA is clocked orders of magnitude faster than the fastest rate of feedback from the encoder. However, the weakness lies in the precision of measurement, with a minimum step

$$\Delta n = \frac{f_s}{N_{enc}}. \tag{8.11}$$

This is unfortunate as the only way to improve precision through changing process parameters is by decreasing the sample rate. Suppose the encoder were sampled at 256 Hz. The quadrature encoder can be used to generate a single signal $A \oplus B$ with double the N_{enc} as either individual channel. Using a 2048 cpt encoder signal, the precision of measurement is then 0.125 or $\frac{1}{8}$ Hz. Already, the sample rate is relatively low and measurement precision is certainly too poor for modulation control.

8.2.2.2 Reciprocal Counter Approach

A *reciprocal counter* simply operates under the reverse approach of the forward counter—rather than counting the number of feedback events which occur during a controller-defined sample period, the controller counts the number of reference clock ticks which occur between two feedback events from the plant. This method is valuable when the forward counter does not provide enough feedback for precise measurement, though it requires the controller to generate reference signals which are significantly faster than the supplied feedback.

Consider a case where the controller generates a clock at $f_{REF} = 200$ MHz with period $T_{REF} = 5$ us. If the controller counts the number of clock ticks N_t between events on the encoder, then the time duration of a full revolution (the period of the motor speed) is the product $T_{ref} N_{enc} N_t$. The frequency of rotation is

$$n[k] = \frac{1}{T_{ref} N_{enc} N_t[k]} \tag{8.12}$$

for the kth observation, hence the measured variable is in reciprocal units as compared with the counter.

Now the maximum sample rate becomes $N_{enc} n[k]$, and there is an abundance of feedback available even at extremely slow speeds. That is, 1024 Sa/s at 1 rev/s is already 4× faster than the modest forward counter. Step resolution is no longer linear, instead we have

$$\Delta n_{min} = \frac{1}{T_{ref} N_{enc} (N_t[k] \pm 1)^2} \approx T_{ref} N_{enc} (N_t[k])^2 \tag{8.13}$$

which holds true for all but the smallest $N_t[k]$ which are inconsequential. More importantly, the value of Δn_{min} is now moderately decoupled from N_{enc} as it is possible to generate extremely fast clocks for which $T_{ref} \ll N_{enc}$, hence, we are free to choose a significantly lower step resolution.

Using a 200 MHz clock having period 5 μs for $Tref$, we find the step resolution is up to 100× greater than for the previously described forward counter, as shown in Fig. 8.9. The sample rate for a reciprocal counter is $\propto n[k]$ and is shown for several normalized values in the figure, all of which yield real-time values > 256 Sa/s for $f > 8$ Hz. The specified sample intervals are well below the maximum value as the amount of feedback is otherwise prohibitively large.

Reciprocal counters come with a large quantity of feedback which also becomes a slight drawback. We have not yet answered the question of how the FPGA will implement the math operation of Eq. 8.12. Multiplication on the FPGA is associated with large resource consumption and time delays, increasing with register width. The frequency step resolution of Fig. 8.9 and corresponding dynamic range shown in Fig. 8.10 create a new design trade-off. We can improve step resolution by decreasing T_{ref}, but each division of T_{ref} by 2 requires addition of another bit to the counter for N_t if the dynamic range is to remain the same. In terms of frequency, there is a diminishing improvement in dynamic range for each additional bit on the counter.

The design floor of measurement precision is chosen here for the sake of example to be 0.0625 Hz or $1/2^4$. A 14-bit register can store fixed-width integer values using 10 decimal bits and 4 fractional bits to cover the integer range 0–1023 with 4-bit decimal precision. If the transmitter is to be used in a channel of 300 Hz, then a counter of this design at a

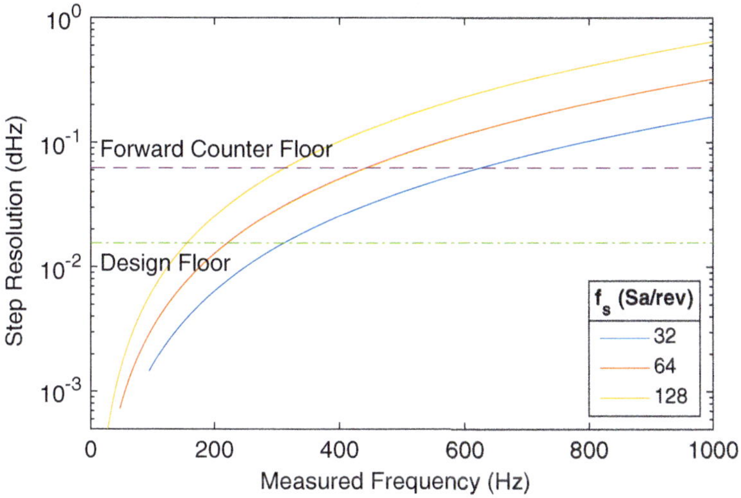

Fig. 8.9 Step resolution of a reciprocal counter with $T_{ref} = 5$ ns and $N_t \in [0, 2^{16}]$ for several values of normalized f_s. The forward counter floor is shown for a realistic best-case $f_s = 256$ Sa/s and $N_{enc} = 4096$ ct/rev. Design floor is shown at $\Delta = 0.0625$ Hz

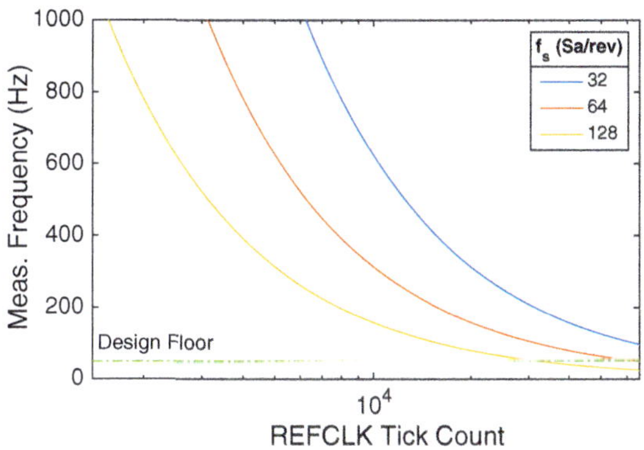

Fig. 8.10 Dynamic range of a reciprocal counter for $T_{ref} = 5$ ns and $N_t \in [0, 2^{16}]$. The design floor is the minimum desired measurement frequency

sample rate of 32 Sa/rev (the blue curve in Fig. 8.9) maintains the desired precision and has a dynamic range of 100–300 Hz. In a channel of 200 Hz, the dynamic range shifts to 50–225 Hz. The sample interval over one full revolution of the motor is configured by routing the quadrature signal $A \oplus B$ through a ripple counter with multiple output taps for reconfiguration on-the-fly.

8.2.2.3 Feedback Near 60 Hz

The transmitter will not be used at 60 Hz, though measurement of speeds near 60 Hz is useful when gathering data (such as the step response of the plant) and for soft-startup to avoid accelerating the load too quickly from rest (potentially causing large transients in the

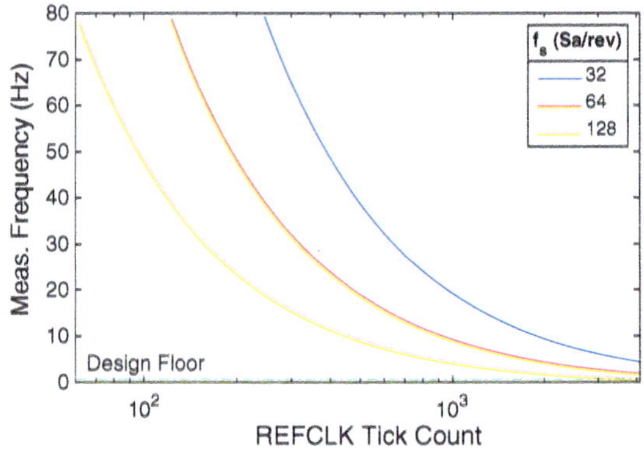

Fig. 8.11 Increasing the reference counter period makes a reciprocal counter useful for feedback near 0 Hz with a workable dynamic range. Values are shown for $T_{ref} = 1.6$ us and $N_t \in [0, 2^{12}]$

power amplifier). Variance in a forward counter renders it ineffective at such low frequencies and the reciprocal counter for ordinary operation is not designed to register values below 50 Hz. There is no need for step resolution below $\Delta n[k] = 1$ Hz, and so we can use a slow reciprocal counter.

Using a 12-bit counter and a polling rate of 64 Sa/rev supplies values with a dynamic range of 5–78 Hz as shown in Fig. 8.11, which maps to a 10-bit fixed-width register using 6 decimal bits and 4 fractional bits. This fallback counter is used by the speed measurement module when measured motor speed (as a rolling average over 8 samples) falls below the minimum registerable value on the standard counter (50 Hz). The step resolution of this fallback counter is summarized in Fig. 8.12. The module switches back to the standard counter when the fallback counter rises to 60 Hz to add stability at the interface. Combinational logic defines the operating boundaries for multiple conditions without adding propagation delay to the counter.

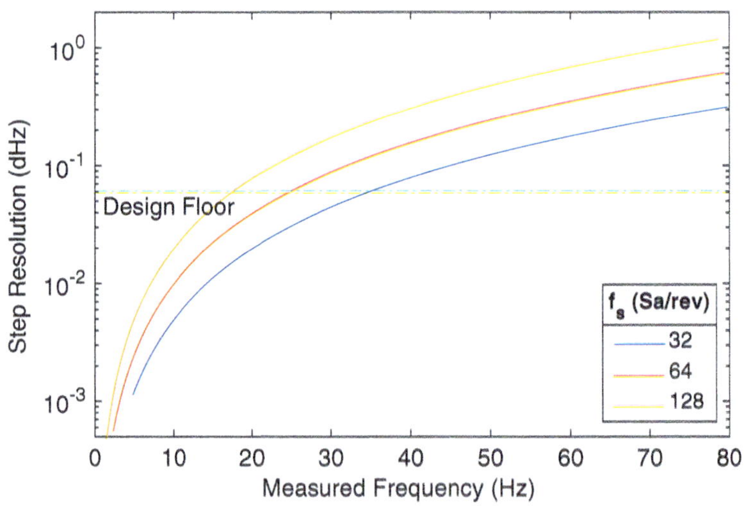

Fig. 8.12 Step resolution of the fallback counter corresponding to dynamic range of Fig. 8.11

8.2.3 Data Collection

A real-time data acquisition module is implemented on the FPGA to measure plant characteristics. The module is controlled and triggered by signals from a parallel communication bus with the HPS which set the desired experimental parameters. In Fig. 8.13, this is described as the characteristic analyzer. The module interrupts the action of the controller to insert changes in manipulated variables while simultaneously recording output from the speed sensor to an incremented address in the off-chip SDRAM. A separate SDRAM controller module presents two separate interfaces, one for data logging from the characteristic analyzer and another for data readout to software running on the HPS.

The SDRAM controller of Fig. 8.13 is a Verilog module which operates at the hardware level to initialize and refresh the off-chip RAM; this is the process of periodically reading and rewriting each bit to maintain the charge on each memory cell. Furthermore, it implements separate read and write ports connected through the top-level design which abstracts the data storage function of the RAM from the instruction set and timing of its IO. The off-chip RAM is clocked at 100 MHz and requires two reference clocks (one which includes a phase delay) which are generated using a phase-locked loop (PLL) intellectual property (IP) core available from the manufacturer (Intel FPGA IP v18.1). The process of data exchange between the SDRAM controller and other modules is the center of a dangerous crossroad—one where three different clock domains must communicate.

Let us consider separately the interface present at the SDRAM controller's read and write ports. *At the read port*, a C program running on the HPS is interfaced through a PIO bus. While the PIO is synchronized to the 50 MHz clock used by the top-level design, the HPS makes requests asynchronously. The DRAM also requires data to be read within a strict time frame which the HPS cannot satisfy. *At the write port*, the plant response is orders of magnitude slower than FPGA fabric. The characteristic analyzer collects data at a software-configurable rate of 0.256–4.096 kHz synchronized to the controller's feedback clock.

The read and write ports of the RAM are fully buffered using first-in first-out (FIFO) Intel IP blocks. HPS read requests are safely carried to the DRAM controller which retrieves the request address from the buffer and places data from the DRAM in the read-out FIFO. The FIFO IP block includes an output to flag when the FIFO is empty. Software on the HPS reads data from the buffer when the FIFO is not empty, then sends an acknowledgment to clear the frame. The next read request is not sent until the previous has already been read to avoid filling the buffers. The write port operates on a similar principle.

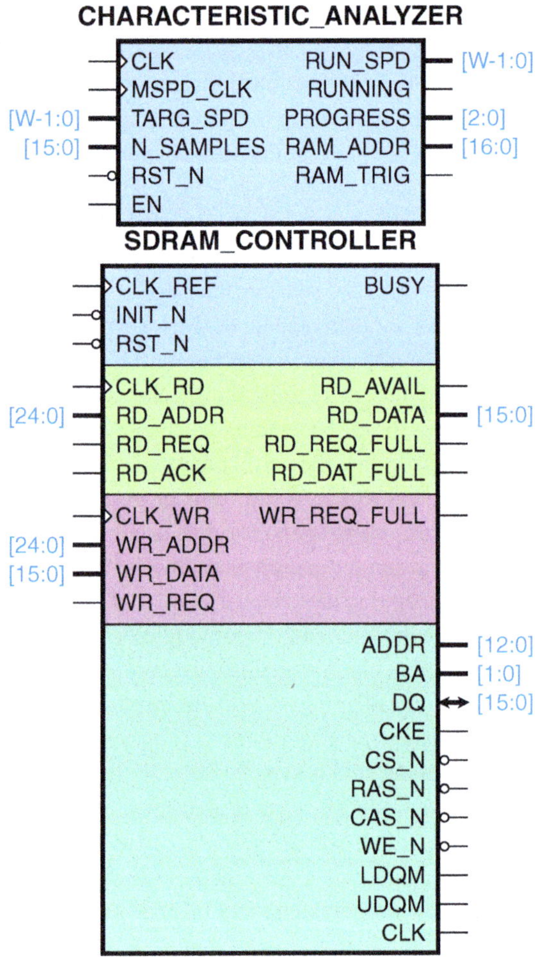

Fig. 8.13 Verilog instantiation blocks for the on-board data recorder and companion SDRAM controller. The SDRAM controller handles clock domain crossing separately for read and write ports

8.2.4 A Model for the Controller

The classic proportional-integral-derivative (PID) controller has frequency response $G(s)$ given by

$$G(s) = K_p + \frac{K_i}{s} + K_d s \tag{8.14}$$

and acts on $e(t)$, the error term $r(t) - c(t)$ between target and measured values of a given plant parameter. Constants[3] K_p, K_i, and K_d are tuning parameters which weight the relative

[3] In some systems, these so-called constants are varied based on set point and/or during ramp-up and ramp-down routines. For the sake of this analysis, we will treat them as constants.

contribution of present, historical, and anticipated future values of system output error to the output $u(t)$.

In the time domain, the corresponding feed-forward controller output $u(t)$ is given by

$$\begin{aligned} u(t) &= K_p e(t) + K_i \int_0^t e(\tau)d\tau + K_d \frac{d}{dt} e(t) \\ &= K\left(e(t) + \frac{1}{T_i}\int_0^t e(\tau)d\tau + T_d \frac{d}{dt} e(t)\right). \end{aligned} \tag{8.15}$$

Note that here τ is an integration variable and not a time constant. The proportional term K_p drives the controller output proportional to the currently observed error. Used alone, proportional control results in steady state error as controller output drops to zero as the plant nears the target output. This error is corrected by the integral term, which applies gain proportional to accumulated error as the system operates. Integral control also improves transient response by providing additional gain when error accumulates over time. However, integral control also tends to introduce overshoot as the system continues to respond to accumulated error as the set point is approached.

The derivative term predicts future behavior, applying gain proportional to the rate of change of the system output. A rapid movement toward the set point results in a large negative $\frac{d}{dt}e(t)$ which reduces overshoot and improves settling time as the system approaches the target. Operating at the set point, rapid movement away from the set point produces a large positive change in error. System instability may be caused by derivative control when system error increases dramatically, known as "derivative kick".

Furthermore, a positive spike in error can cause the derivative term to apply gain which amplifies a trajectory away from the set point. For a rotating mechanical system, consider a case where the driven load undergoes a transient acceleration such as when the mass imbalance of the rotating shaft is aligned vertically and gravitational acceleration affects the system. The derivative term amplifies this acceleration by increasing $u(t)$ in response. In the alternative case when the system is slowed by transient forces producing a drag, the derivative term is quite useful to quickly correct transients as they occur.

PID control has been in use for about a century, with many existing examples using analog controllers. Digital controllers have become a subject of recent interest due to their low cost of implementation and the relative ease of software-based adjustment [1, 2]. Fast, parallel, reconfigurable digital logic enables the FPGA to implement complex application-specific controllers which can be adapted in response to changes in process parameters or the plant [3].

Efficient implementation of PID control on the FPGA requires some adaptation to hardware-level design. All operations must be carried out at discrete time intervals, use minimum system memory, and make efficient use of logic elements. Equation 8.15 is rewritten in discrete time to meet these ends, whereby the present output from the controller u_n is

$$u_n = K_p e_n + K_i T_s \sum_{i=0}^{n} e_i + \frac{K_d}{T_s}(e_n - e_{n-1}) \tag{8.16}$$

where T_s is the sample period. Memory in the integral term remains a challenge. While it is common in practice to regard the accumulated error $\sum_{i=0}^{n} e_i$ as a representation over some finite time interval, even an 8 s history at a sample rate 256 Hz requires the storage of 2048 17-bit values. Furthermore, the sum of these values requires a large number of sum operations, introducing latency and consuming device space. An alternative is to implement a recursive algorithm by first calculating Δu

$$\Delta u_n = u_n - u_{n-1} = (K_p + K_i T_s + \frac{K_d}{T_s})e_n - (K_p + \frac{2K_d}{T_s})e_{n-1} + \frac{K_d}{T_s}e_{n-2} \tag{8.17}$$

then isolating u_n and defining terse coefficients which yield

$$u_n = u_{n-1} + K_0 e_n + K_1 e_{n-1} + K_2 e_{n-2} \tag{8.18}$$

where

$$\begin{aligned} K_0 &= K_p + K_i T_s + \frac{K_d}{T_s}, \\ K_1 &= K_p - \frac{2K_d}{T_s}, \text{ and} \\ K_2 &= \frac{K_d}{T_s}. \end{aligned} \tag{8.19}$$

Each time Eq. 8.18 is evaluated, the new value of u_n depends only on its previous value u_{n-1} and the three most recent observations of e_n. The computation requires only three multiplication and three addition operations after the computation of e_n.

Implementation of the recursive algorithm in the FPGA fabric is realized by the block diagram of Fig. 8.14. The design uses three multiplication operations, four addition operations, and five data registers. Setup and hold times for math operations are satisfied by a clocked delay line in simulation. Update operations are triggered by the positive edge of a PID Clock at frequency f_{PID} which also uniformly samples the target and feedback data. Delay line blocks operate at a minimum frequency $10 \times f_{PID}$ in the simulation.

Hardware implementation uses a Moore machine to realize behavior equivalent to Eq. 8.18, using the positive edge of f_{PID} as an update trigger and inserting a small (1–5) number of clock cycles on the top level 50 MHz clock between fan-out operations based on requirements of each stage. Saturation control at the output serves to (i) prevent controller overflow and (ii) allow reduction of maximum drive duty below 100% for the safety of the plant. Figure 8.15 shows the gate-level logic model for such a controller, where CLK is the top-level design clock (50 MHz). Enable signals L1–L3 are used to pipeline math operations and ensure timing requirements are satisfied.

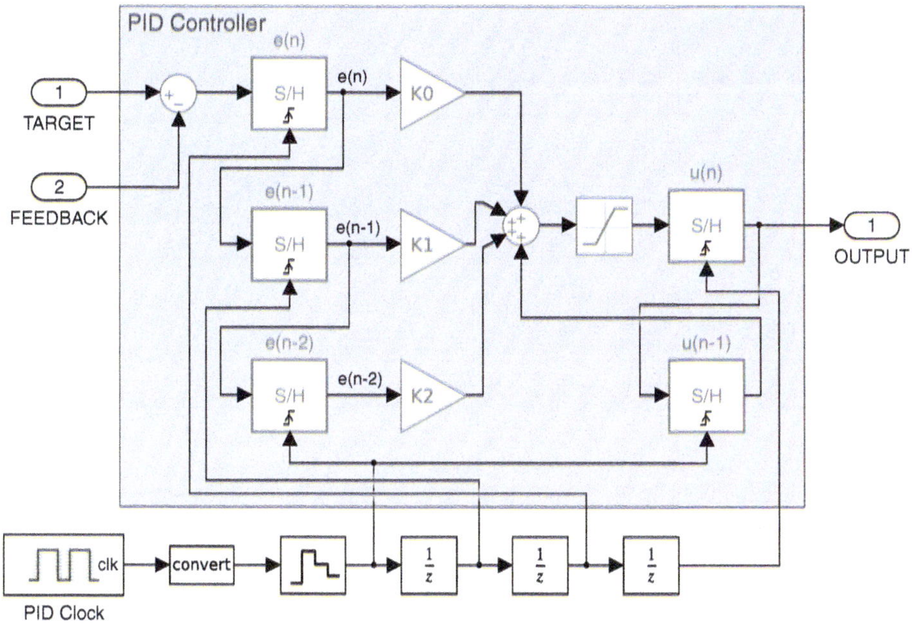

Fig. 8.14 Simulink model of the discrete recursive PID control algorithm

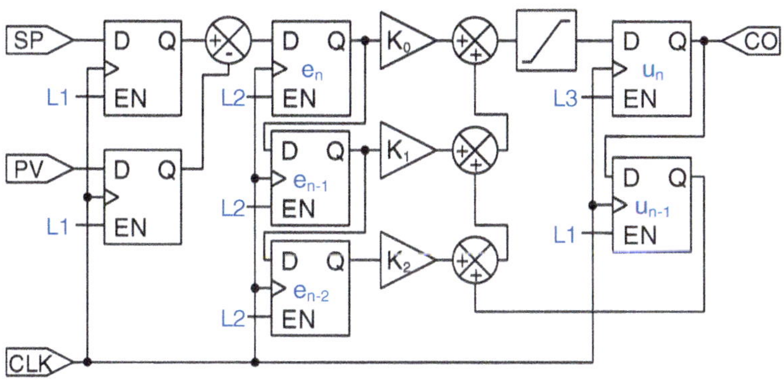

Fig. 8.15 Gate logic equivalent synthesis of PID control algorithm

Alternatively, it is somewhat less complex to design the hardware module around the classical PID control structure as shown in Fig. 8.16 by using a register-based accumulator to compute the integral term. In a well-behaved system with appropriately large memory allocation for registers Integral and Sum, the implementation is equivalent to the ideal response. In the case that the dynamical system deviates from such good behavior, the hardware operation of updating the value in each register is gated by a logical comparison

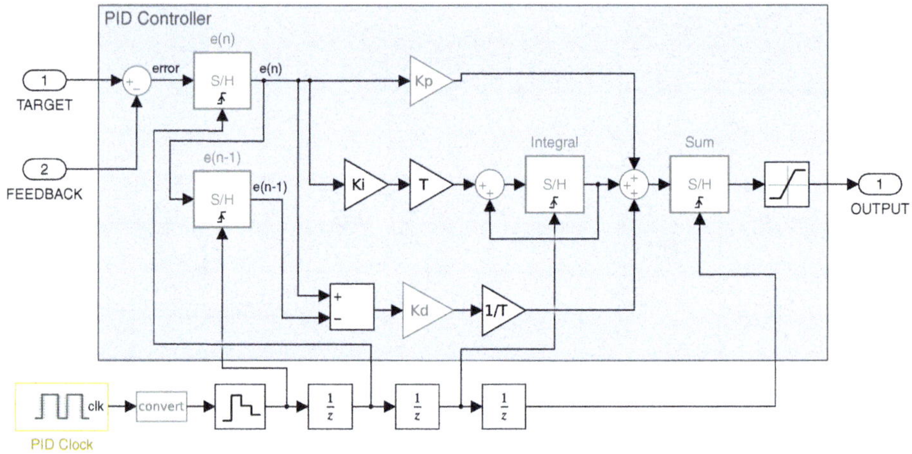

Fig. 8.16 Simulink model of discrete classical PID control algorithm optimized for hardware logic synthesis

which detects the condition and blocks the register update (this behavior is not shown in Fig. 8.16).

This design uses three multiplication operations, five addition operations, and four data registers. The realizable improvement in overall resource utilization and computation efficiency is greater than it appears. Hardware implementation of multiplication blocks (lpm_mult Intel FPGA IP v18.1) utilizes DSP blocks with propagation delay increasing with register size.[4] Computation time is also larger for floating point operations. Values of $K0$, $K1$, and $K2$ from Eq. 8.19 tend to result in significantly larger multiplication blocks, for example (at a sample rate $f_{PID} = 4096$ Hz) if $K_p = 5$, $K_i = 200$, and $K_d = 0.15$, the corresponding values are $K0 = 66.5$, $K1 = -128$, and $K2 = 61.4$. Using constants K_p, K_i, and K_d directly is more efficient to implement and allows the constants to be normalized by the sample rate and/or converted between fractional and fixed-width representation on demand.

8.3 Software Interface

System testing and operational functions are implemented as a set of console applications on the HPS, shown inside the Linux operating system (OS) of Fig. 8.17. The command-line interface (CLI), also called the AMEBA2 Console, is a C implementation which runs in a Bash shell to map FPGA fabric memory and directly configure hardware PIO buses.

[4] This is a stepped function of register size, increasing as the number of DSP blocks for a single operation increases.

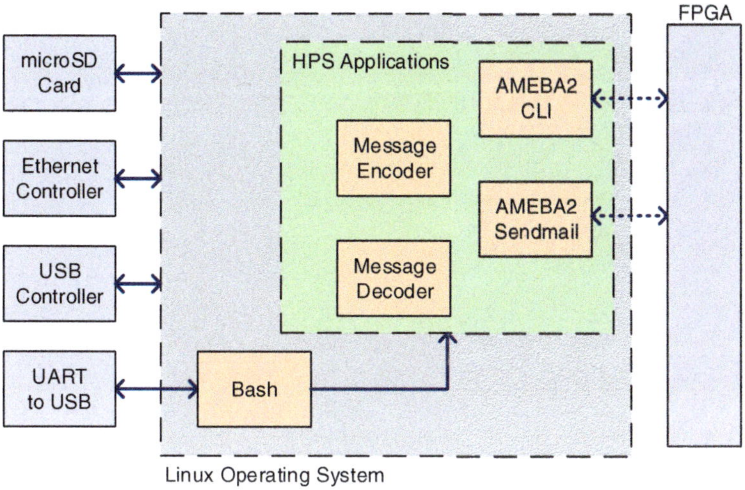

Fig. 8.17 Organization of user-level software components

AMEBA2 Sendmail is an automated program which reads from user input or defaults to a text file, then loads and transmits packets on user command. AMEBA2 Sendmail uses interface libraries provided by the source tree for the development of the console, thus is essentially a scripted version of console commands with additional high-level control. Message encoder and decoder applications operate on either text files or direct user input and apply the series of operations defined in the script.

The console operates in low-level ("bit-banging" style) and high-level (control performance evaluation) modes and uses pre-defined code routines to ensure incorrect sequences are not accidentally passed to the power amplifier. The controller is capable of self-debugging and plant performance logging. Pre-defined code routines configure FPGA fabric and execute a series of operations to divert process variables and plant feedback into a RAM. After an experiment is complete, the data are written to text files with a header including collected metrics and configuration parameters. Real-time data logging is achieved using a sample clock configured for 256–4096 Hz through ripple clock taps.

All experiments presented on plant and controller operations make use of these built in data logging features to demonstrate the functionality of the system and analyze transmitter performance. For example, Fig. 8.18 shows the console application being used to record an open-loop step response. Open-loop control directly specifies linear values between 0 and $2^{14}-1$ which map to duty cycle 0–100% for the commutating PWM drive signals.

Using the step response measurement mode, the open-loop response of the plant to unit step functions can be measured for use in tuning the PID module. Figure 8.19 plots a sequence of four step response measurements for acceleration and deceleration between 110 and 150 Hz.

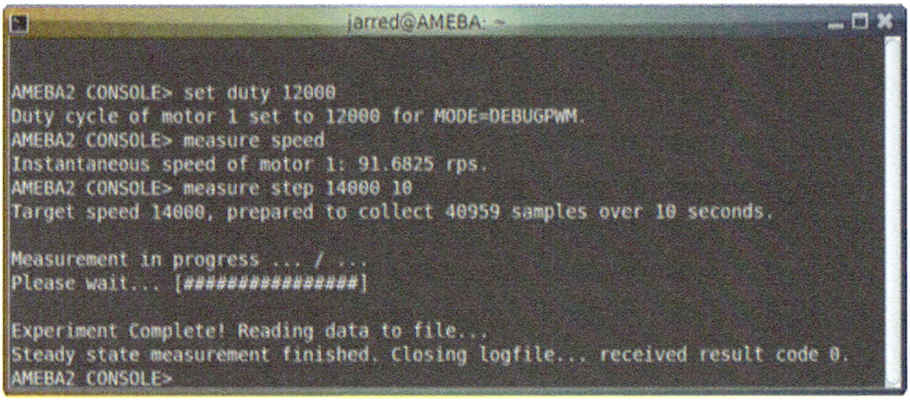

Fig. 8.18 Console screenshot after measuring step response

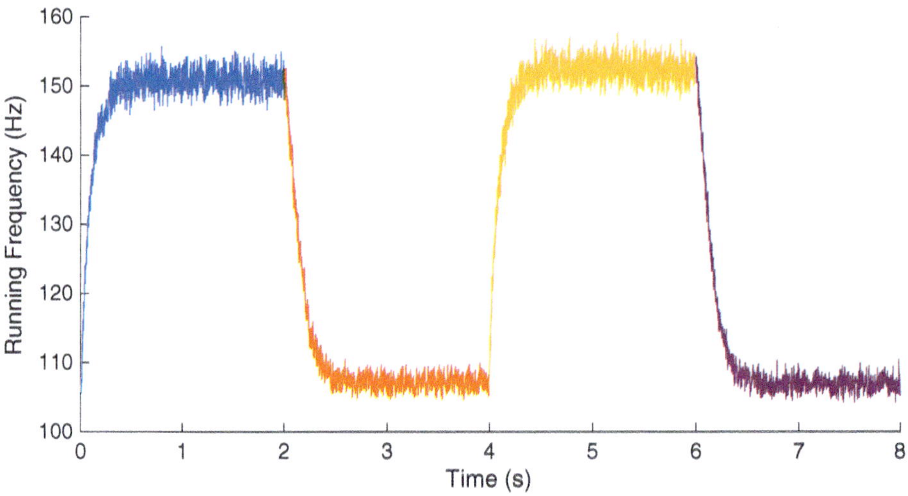

Fig. 8.19 Step response measurement using the debug console

8.3.1 Datatype Organization

Messages are transmitted through the channel as a collection of packets, each containing eight symbols preceded by a parity signal. While the parity signal is identical in width and dynamic range to a packet, for convenience it may take structures beyond the level transitions of the modulation dictionary. This allows the parity symbol to convey additional information, for example, consider two distinct "unit caret" shapes ^ and ᵥ used as the parity symbol. The symbols convey binary "even" or "odd" designation and produce a sequencing structure useful for detection and windowing in the receiver over long messages (or in the event of a delay during transmission).

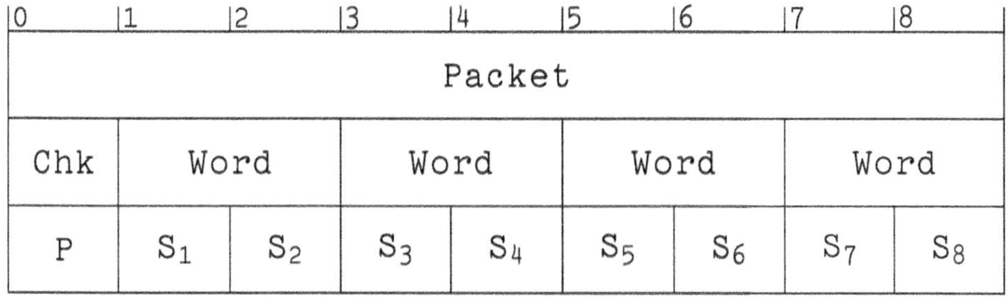

Fig. 8.20 Data types defined for communication link: Packets contain 9 symbols organized into a parity/control signal (which is the size of one symbol) followed by four words. Each word contains two symbols and represents one character

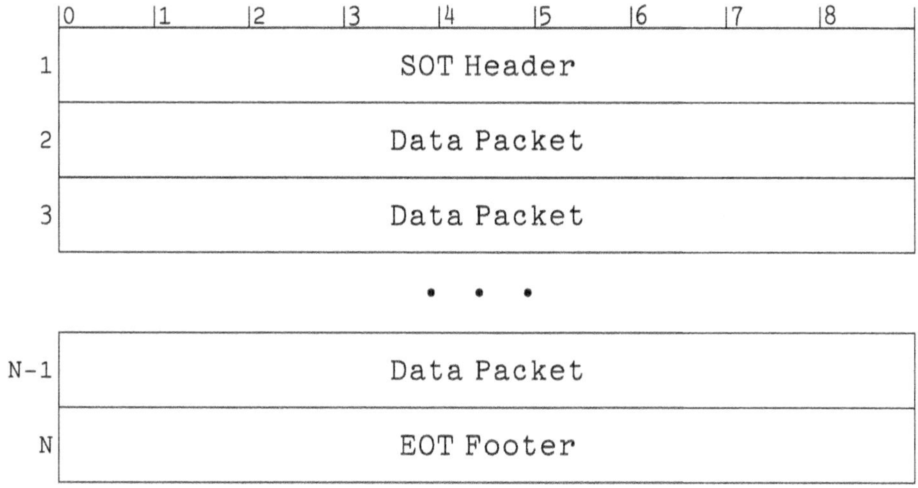

Fig. 8.21 Message content is organized into packets with `SOT` and `EOT` frames used for signaling

Figure 8.20 enumerates the organization of symbols into a packet as described. A single ASCII character is conveyed in two symbols and thus each pair is also grouped into a word. ASCII characters and control sequences are mapped to defined `words` using a dictionary and this map may be re-defined by replacing the dictionary as needed, in effort to reduce the random walk effect observed in Sect. 5.3.7. We begin by mapping the standard alphabet which parallels the numeric transform of Sect. 4.3.6.

A full message is organized into a series of data packets. Before they are transmitted, messages are encapsulated by header and footer packets as in Fig. 8.21 to indicate the start and end of a message, respectively. Additional error or validation checking could also be implemented in the control packets (such as a message checksum) though this is reserved for future work.

8.3.2 Message Encoding

The message encoder is a software component (an application in Fig. 8.17) which operates on the message to produce a baseband signal. Compression utilizes the same scheme as the prototype in Sect. 5.16. After compression, characters are mapped to words and grouped into packets. Since the controller output is the linear value $\omega[t]$, upconversion to the carrier frequency is a simple matter of addition. Thus, packets can be modulated at the baseband in advance of transmission.

The modified encoding scheme is provided in Fig. 8.22. The baseband and upconverted signals are discrete vectors which contain the complete transmit signal formatted in the fixed-width representation required by the FPGA controller.

8.3.3 Sendmail: Real-Time Data Modulation

The sample rate must match that of the feed-forward PID controller, which is minimally sized to maintain efficient use of the FPGA fabric for the packet sender RAM and minimize accumulation error. Regarding the PID controller, the sample rate must be fast enough that the controller is sensitive to all dynamics of the plant [3, 4]. The absolute requirement is subject to various guidelines, and it is possible to sample as fast as feedback is made available. In the literature, discrete time PID controllers are reported to be extremely sensitive to slight variation in sample time [4]. Hardware resource utilization must also be increased to prevent the accumulation of discrete computation errors.

The phase trajectory for a set of 11 data packets is shown in Fig. 8.23. The information source is the message,

ALL THINGS EXIST AS THEY ARE PERCEIVED...

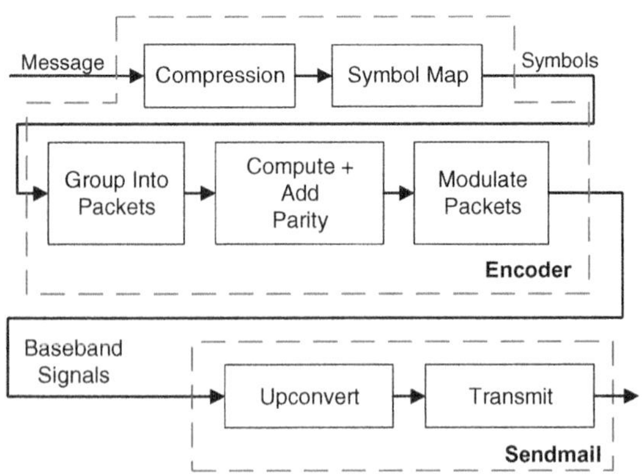

Fig. 8.22 Encoding and transmission scheme for complete message

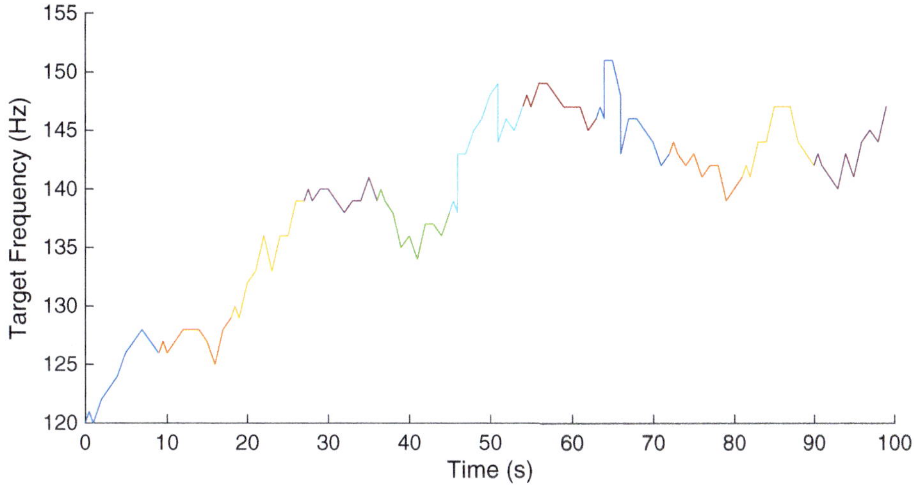

Fig. 8.23 Sequence of 11 packets produced by the Sendmail program. Each packet is displayed in a unique color

which serves a dual purpose: as example text and also as suggested words of wisdom from the authors.

8.4 Experimental Performance Characteristics

The steady state behavior of the mechanical transmitter is observed over a 10 s sample interval in the open-loop control regime, which is shown in Fig. 8.24. In open-loop control, the manipulated variable is the duty cycle, and actual speed varies from what is expected based on calibration mapping. In the figure, the measured steady state output mean is displayed above each of the six subfigures.

Open-loop performance was now evaluated. Data was used to tune the PID controller. The closed-loop PID controller was then implemented and performance compared to the reference data. Some reference steady state speed measurements are shown in Fig. 8.24 in the open-loop case. The plant has been operating for several minutes. Long-term effects of localized heating are not considered in this reference case, and the transmitter is operated as though it were in use (where it likely starts from rest).

8.4.1 Testing the Drive

Using a PWM carrier at 195 kHz, the data in Fig. 8.25 were collected from a maxon ECX16L motor and a supply voltage 24 V. The PWM duty cycle was incremented and speed measured

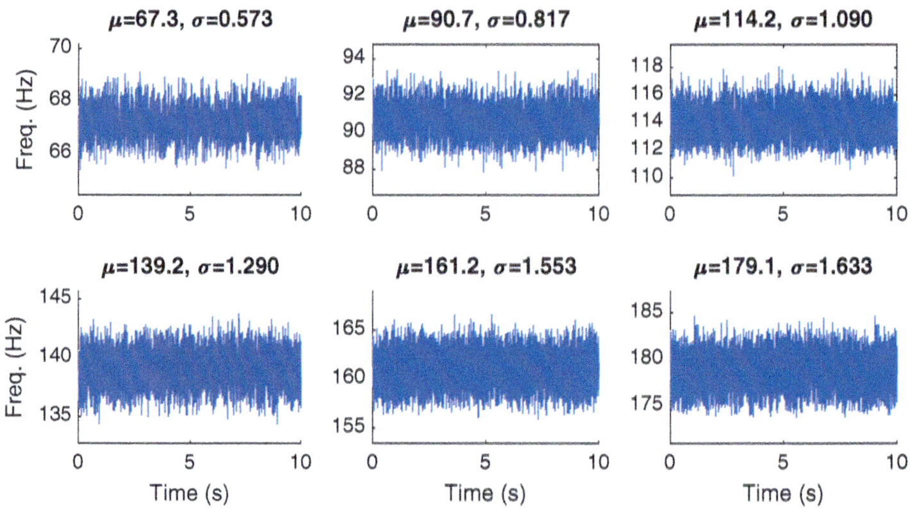

Fig. 8.24 Steady state behavior of the transmitter in open-loop control regime

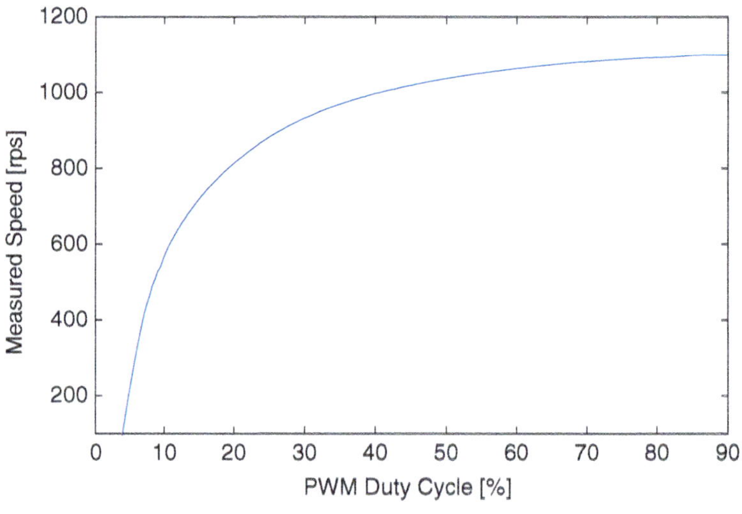

Fig. 8.25 Observed speed of an unloaded BLDC motor at varied input duty cycle, $f_{PWM} = 195$ kHz

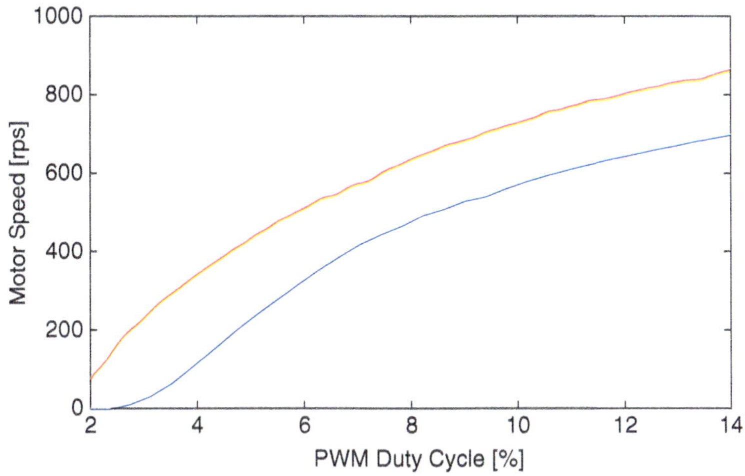

Fig. 8.26 Comparison of measured speed in response to varied duty cycle using several PWM frequencies for a maxon ECX16L motor; blue: $f_{PWM} = 195$ kHz, red: $f_{PWM} = 48.8$ kHz

after several seconds at each point along the curve. The response of the motor is characteristic of its torque-speed curve, with the exaggerated trend owed to drive switching delays and conservative gate drive current limits.

The same experiment is repeated with f_{PWM} decreased to 48.8 kHz. Figure 8.26 shows a comparison of this response to data from the previous experiment. Lowering the PWM frequency reduces the total switching time by a factor of $4\times$, causing the constant shift in the speed response. Nonlinear behavior at low PWM on time is also reduced, as this corresponds with the minimum duty cycle required for complete switching of the power amplifier FETs.

The output resolution of the PWM signal generator is the change in motor speed in response to a unit adjustment to PWM "on" time. Output resolution is measured by repeating the same experimental procedure using an 8-bit counter (precision 0.39%) which is then increased to 12 bits (precision 0.024%). Figure 8.27 compares the output resolution and demonstrates the increased drive resolution available. There is, however, a limit to the achievable output resolution based on the maximum possible reference clock. The highest speed grade Cyclone V SoC has a maximum clock speed of 550 MHz. Using 500 MHz provides a safe ceiling for reliable long-term operation and yields a switching speed of $f_{PWM} = 30.5$ kHz for a 14-bit counter. The marginal increase afforded by a 550 MHz clock is not worth the possible hassle of operation at the specified limit.

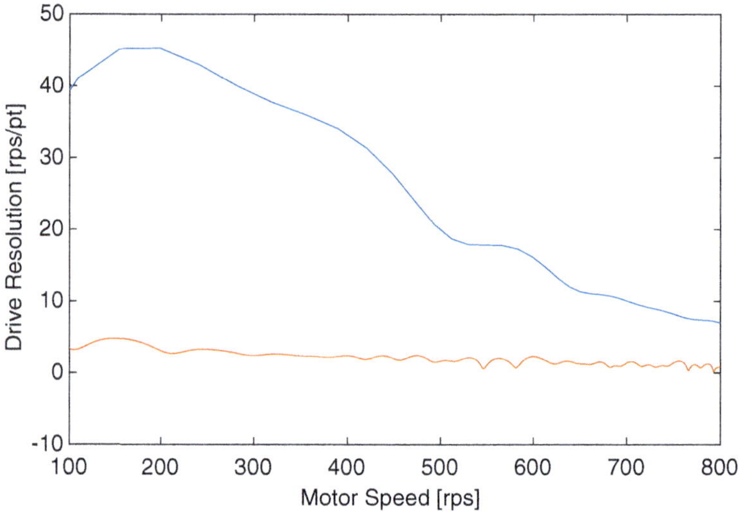

Fig. 8.27 Measured impact of PWM generator resolution on motor speed modulation resolution for a maxon ECX16L motor; blue: 8-bit counter, red: 12-bit counter

8.4.2 Tuning the Controller

The PID controller is tuned using the step response of the real plant performed in comparison to a model. It is convenient for the sake of controller tuning to consider the plant as a first-order system plus dead time (FOPDT). This system has been well studied in the literature since the famous work by Ziegler and Nichols in [5]. Thus, we will consider the model

$$G_c = K \frac{1}{Ts + 1} e^{-Zs}, \tag{8.20}$$

where Z denotes the dead time.

We will soon see that a first-order model $K/(Ts + 1)$ with a single time constant T is a suitable approximation to Eq. 8.6 for the plant. Dead time due to cumulative delays in the controller and driver is described by the time delay term e^{-Zs} (where Z is the time delay). Such a model has been utilized for tuning and control of unstable processes and nonlinear systems [6, 7]. In any case, a controller with adaptive gain adjustment for on-the-fly coverage of multiple operation modes has also been shown [8].

Figure 8.28 shows open-loop step response behavior of the plant measured for various amplitudes of excitation. Units for excitation are still in fractional D and the appropriate value for a given expected speed is based on a map such as that of Fig. 8.26. Regarding the measurement method, there exists a short delay at $t = 0$ before the motor rotates fast enough for the speed sensor to report values above zero. Low (< 10) speed values in some runs are erroneously recorded with nonzero start speed due to the switchover threshold for

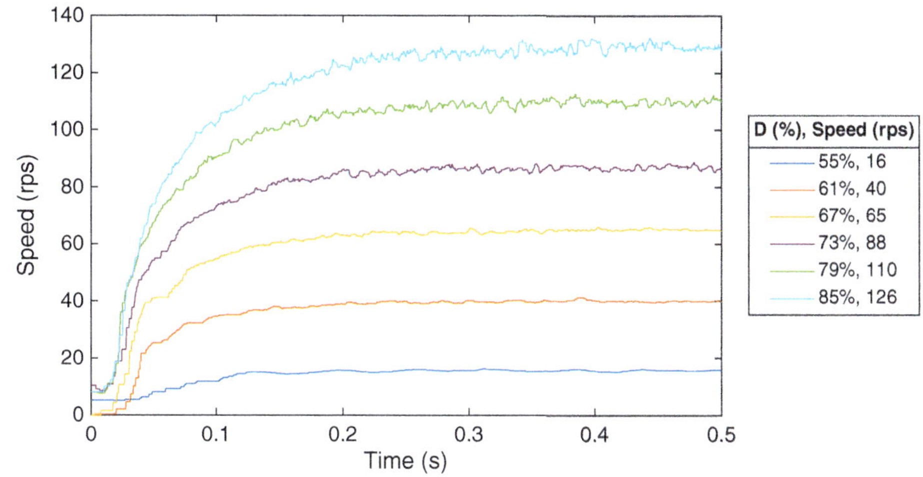

Fig. 8.28 Open-loop step response for various excitation amplitudes. The transmitter starts in the rest state

the fallback counter and the use of only synchronous resets in the particular version of the speed sensor used to collect the data.

Observing a single measurement curve, the data are a good fit to the FOPDT model $e^{-Zs}\frac{1}{Ts+1}$. In Fig. 8.29, the model values $Z = 0.005$ sec and $T = 0.055$ sec are used. Dead time is due to cumulative delays in the controller and driver, separated from the plant response with a time constant T. We can observe that the FOPDT model is a good fit for the experimental data, allowing us to reasonably approximate the plant using such a first-order model. Thus, based on the step response Ziegler and Nichols would recommend a PI controller with $K_p = 9.9$, $K_i = 600$, and $K_d = 0$. If including the derivative term, their recommendation would be $K_p = 13.2$, $K_i = 1000$, and $K_d = 0.033$.

Fine adjustment is accomplished using software tuning in Simulink, a useful tool to quickly tweak the balance between controller response time and stability. Simulated step response using the PID controller is compared in the open-loop case to the untuned response in Fig. 8.29. Here, the FOPDT model is representative of the plant, though the computed second-order response is also noted to be an equivalent fit. It is critical that empirical data is collected across the full operating bandwidth of the transmitter, to confirm that the first-order model (which does not capture frequency-dependent effects of drag, commutation switching speed, etc.) continues to hold.

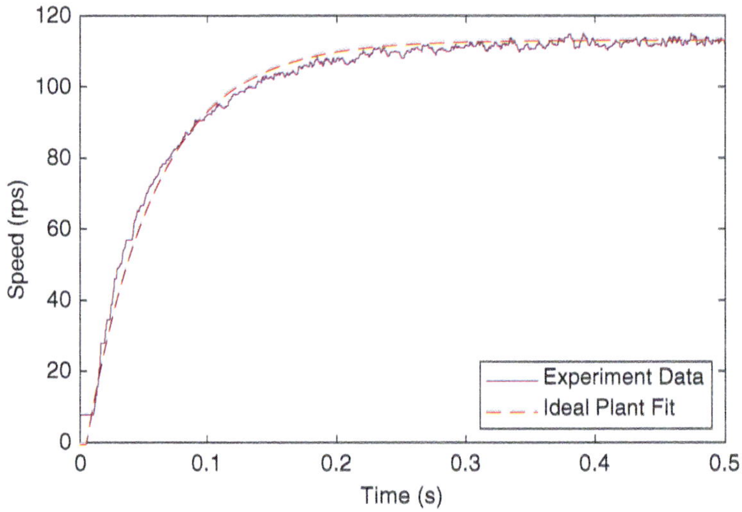

Fig. 8.29 Recorded open-loop step response and fit to first-order system with loop delay

8.4.3 System Response and Stability Margins

The complete transmitter system has a continuous time feed-forward transfer function

$$G(s) = K \cdot G_c(s) \cdot H_m(s) = e^{-Zs}\frac{K_p + \frac{1}{T_i s} + T_d s}{Ts + 1}, \tag{8.21}$$

where both controller and plant are modeled using their ideal continuous time response established in Sects. 8.1.2 and 8.2.4. Here, the discrete nature of each is captured by the single delay with time constant $Z \approx 0.005$ s found previously by empirical analysis. We note (as previously anticipated) that the dominant source of propagation delay is by mechanical loading of the plant due to inertia, since it has been well established that all other system components operate considerably faster in comparison with Z.[5] To go a step further, it is well established in Chap. 6 that the torque τ_{mod} imposed by inertia of the permanent magnet and motor rotor contribute to a first-order system response, while air friction is responsible for a second-order effect. Once again, despite its simplicity, the open-loop step response confirms what we have already spent considerable time simulating—the product of our design effort is a realized mechanical $\vec{B}$-field source for which carrier field generation imposes no energy toll. Numerical results of computer simulation and modeling were tabulated in Table 6.1 and agree with this conclusion.

[5] Take for example, the controller (sampling delay 244×10^{-6} s), power amplifier (switching time 710×10^{-9} s), and motor (stator windings with $\tau_e = 554 \times 10^{-6}$ s).

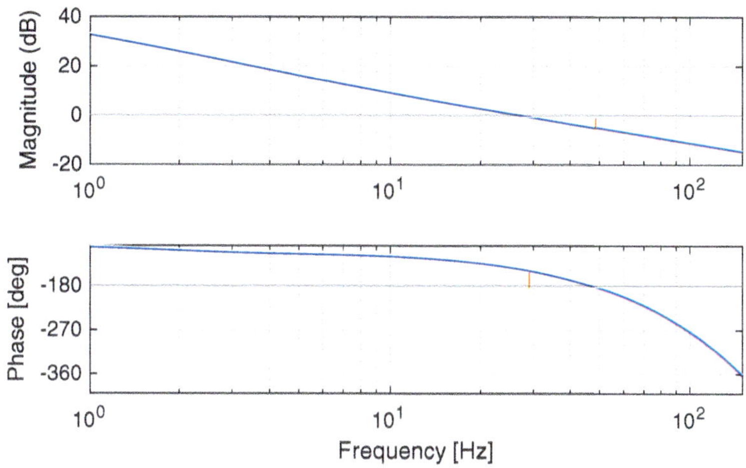

Fig. 8.30 Open-loop frequency response for $K_p = 10$, $K_i = 300$, $K_d = 0$ with stability margins marked at values given in text

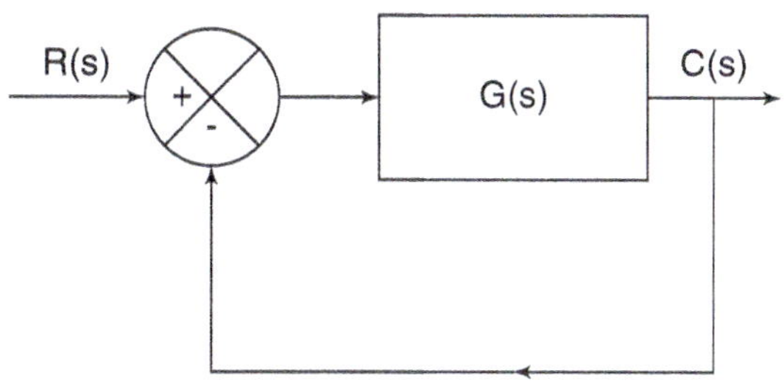

Fig. 8.31 Simple is genius: the transmitter is reduced to a single block response with unity gain feedback

Bode gain and phase plots of Eq. 8.21 are produced in Fig. 8.30 using controller gain values $K_p = 10$, $K_i - 300$, $K_d = 0$. In this configuration, the gain margin is 4.5 dB and occurs at the gain crossover frequency 29 Hz. Similarly, the phase margin is 33.9deg and occurs at phase crossover 40 Hz. Using instead the values $K_p = 10$, $K_i = 1000$, $K_d = 0$, we find (not reflected in Fig. 8.30) the system has gain margin 2.3 dB, gain crossover frequency 32.2 Hz, phase margin 10.9deg, and phase crossover frequency 40.3 Hz. We see that for this system the guidelines of Ziegler and Nichols yield aggressive controller action near the boundary for instability. Hence for improved robustness, the lower integral gain is preferred to maintain system stability.

Now the complete system is described by applying unity gain feedback and closing the loop around the system $G(s)$ from Eq. 8.21. We find the result is quite elegant, arriving at the final block diagram for the system as shown in Fig. 8.31. Magnitude and phase plots of the system frequency response are given in Figs. 8.32 and 8.33, respectively. The loop bandwidth for conservative operation is 64.4 Hz. A comparable bandwidth of 63 Hz is maintained under more aggressive controller action.

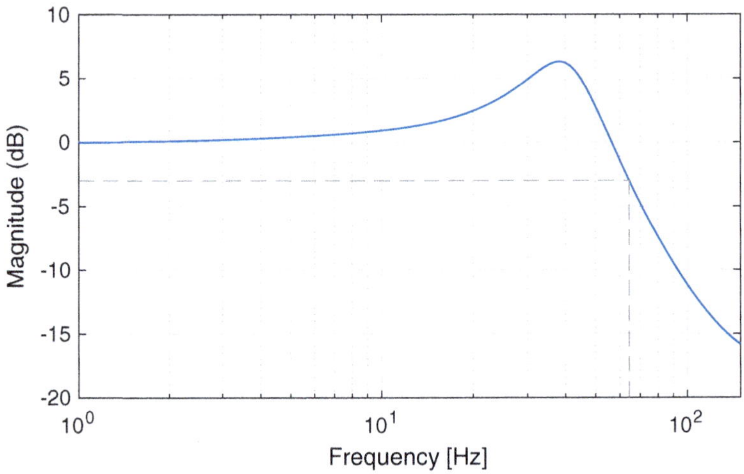

Fig. 8.32 Gain of closed-loop system frequency response for $K_p = 10$, $K_i = 300$, $K_d = 0$

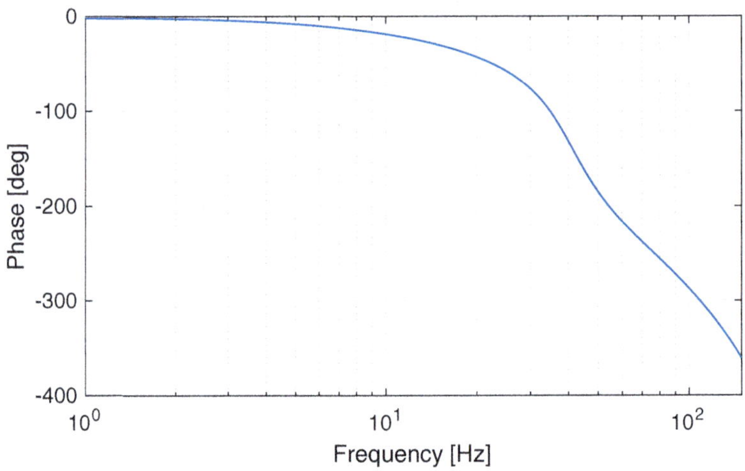

Fig. 8.33 Phase of closed-loop system frequency response for $K_p = 10$, $K_i = 300$, $K_d = 0$

It is possible to adjust the three PID gains on-the-fly using the embedded Linux system on the controller. Selecting several interesting values for K_i, the step response measurements of Fig. 8.34 are produced. The response for $K_i = 300$ is considered as the reference case to yield a well-behaved system, with the typical settling time observed to be less than 100 ms in practice (as shown in the Figure). Now the system is independent of plant parameters and the controller maintains stable control within ~ 2 Hz without the use of derivative action. Applying derivative action can be used to remove ripples periodic with the frequency of operation, though not without the risk of potential system instability. Noting the ripples of

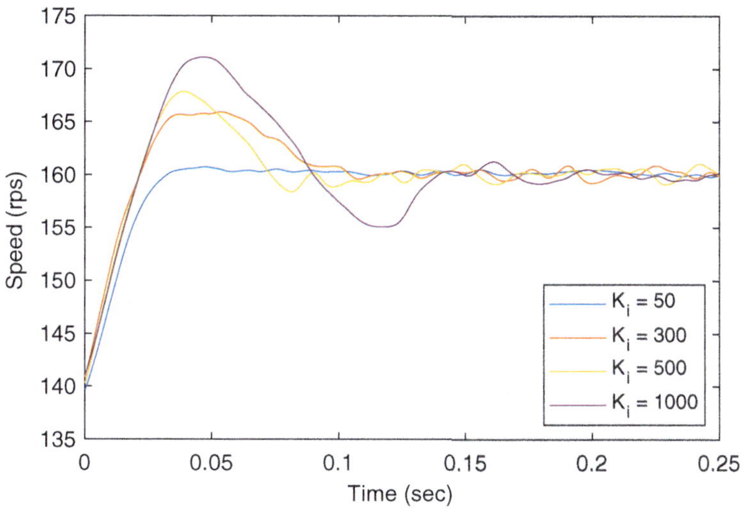

Fig. 8.34 Step response measurement in open loop for varied K_i with constant $K_p = 10$, $K_d = 0$

Fig. 8.34, the two principal causes are a mass imbalance of the rotating body and measurement error periodic with the rate of rotation. The former is the dominant effect, with the latter caused by jitter (nonlinearity and accumulated errors at the point of observation) from the motor encoder. For greater detail, refer to Fig. 7.2, Sect. 7.1.1.

8.4.4 Modulation Energy Explained

Section 5.3.6 gives a comparison of energy consumption for CP-FSK and CF-FSK in the original AMEBA prototype. We find the prototype transmitter requires effectively the same energy to transmit using each of the two modulation schemes, and postulate that the effect is due to massive power losses of drag, friction, etc. AMEBA2 suffers no such dramatic losses.[6] Thus, AMEBA2 allows us to further investigate the claims of modulation energy variations. Recall the definition for a generic M-ary CPM waveform $s(t)$ given by Eq. 4.1 for which the transmission frequency at any value t takes one of M discrete values.[7]

Such a signal may appear roughly as shown in Fig. 8.35, which actually depicts the debug datalog produced by the AMEBA2 Sendmail program during the modulation routine which is run prior to transmission. It is unreasonable to drive the plant with a purely conventional CP waveform since it is by nature confined to a finite bandwidth.[8] The target vectors displayed

[6] Support for this statement is found in rigorous theoretical analysis (particularly that of Sect. 6.3) and furthermore by empirical indications such as in Sect. 8.4.3.

[7] This is somewhat similar to MFSK except with continuous phase at intermediary boundaries between symbols.

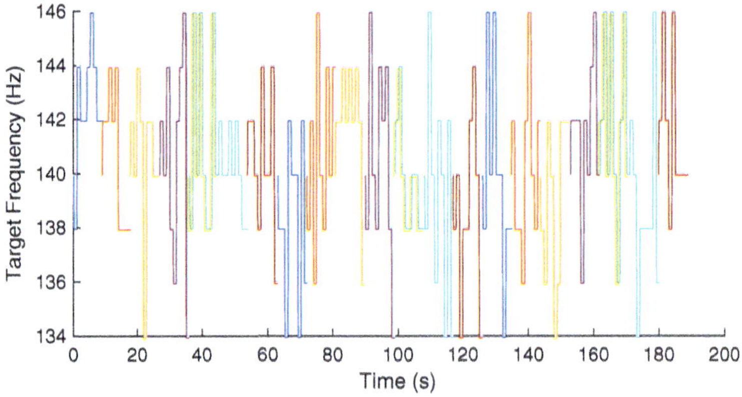

Fig. 8.35 Target frequency visualization over 21 symbols generated using the modified-CP modulation scheme (CP-FSK)

in the figure actually depict a slightly modified version of Eq. 4.1 by including a low-pass filter applied only to a brief windowed portion of the data covering the transitions between symbols. This is our first test case.

Our second test case is the same message modulated using CF-FSK as in Eq. 4.2. The result is a set of modulation vectors as those shown in Fig. 8.36. Comparison between the modulation schemes of Figs. 8.35 and 8.36 is well illustrated by Fig. 4.10.

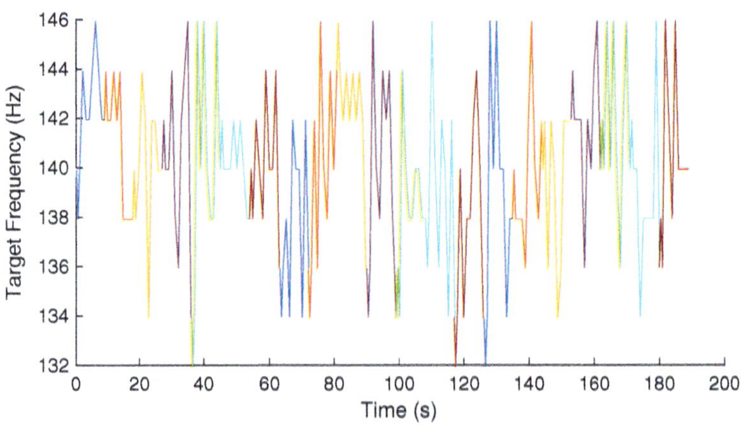

Fig. 8.36 Target frequency visualization over 21 symbols generated using the CF modulation scheme having confined phase (CF-FSK)

[8] Driving the plant with a target behavior which deviates beyond the boundaries of the controller bandwidth also risks destabilizing the system by causing excessive integrator windup.

Each modulated message is transmitted 10 times and energy use measured using the current sense probes on the power amplifier board (Sect. 7.2). We find the message utilizes 3.8 kJ by modified CP-FSK (average power 12.8 W) and 3.7 kJ by CF-FSK (average power 12.5 W, a 2.63% reduction in the latter case).

Since the second generation electromechanical transmitter achieves near-zero static power consumption, the power consumption measurement is expected to be $\approx P_{mod}$. Average transmit power consumption is expected to scale proportional with f_c, so the same modulated message transmission experiment is repeated using 100 Hz carrier frequency to confirm the trend. In this case, the power consumption is 9.0 and 9.1 W for CP-FSK and CF-FSK modulation, respectively, a $\sim 1/1.4\times$ reduction over the previous experiment.

References

1. Subasri, V., Lavanya, K., & Umamaheswari, B. (2006). Implementation of digital PID controller in field programmable gate array (FPGA). In *2006 India International Conference on Power Electronics* (pp. 172–176). IEEE.
2. Kocur, M., Kozak, S., & Dvorscak, B. (2014). Design and implementation of FPGA-digital based PID controller. In *Proceedings of the 2014 15th International Carpathian Control Conference (ICCC)* (pp. 233–236). IEEE.
3. Sathyan, A., Krishnamurthy, M., Milivojevic, N., & Emadi, A. (2009). A low-cost digital control scheme for brushless DC motor drives in domestic applications. In *2009 IEEE International Electric Machines and Drives Conference* (pp. 76–82). IEEE.
4. Laskawski, M., & Wcislik, M. (2016). Sampling rate impact on the tuning of PID controller parameters. *International Journal of Electronics and Telecommunications, 62*(1).
5. Ziegler, J. G., & Nichols, N. B. (1942). Optimum settings for automatic controllers. *Journal of Dynamic Systems Measurement and Control—Transactions of the ASME, 64*(11).
6. Reynoso-Meza, G., Carrillo-Ahumada, J., Boada, Y., & Pic'o, J. (2016). PID controller tuning for unstable processes using a multi-objective optimisation design procedure. In *11th IFAC Symposium on Dynamics and Control of Process Systems Including Biosystems (DYCOPS-CAB 2016)* (Vol. 49, No. 7, pp. 284–289). IFAC-PapersOnLine.
7. Yurkevich, V. D. (2009). Adaptive gain tuning in nonlinear control systems designed via singular perturbation technique. In *2009 IEEE Control Applications, (CCA) & Intelligent Control, (ISIC)* (pp. 37–42). IEEE.
8. Lee, J. Y., Jin, M., & Chang, P. H. (2014). Variable PID gain tuning method using backstepping control with time-delay estimation and nonlinear damping. *IEEE Transactions on Industrial Electronics, 61*(12), 6975–6985.

ELF Receiver Design 9

9.1 Overview of the Receiver

The proposed receiver uses orthogonal coils to sense all three components of the ELF **B**-field. Moving to a 3-axis detector improves communication distance by (i) improving tolerance to transmitter-receiver misalignment, and (ii) increasing the available field over the one-axis case; the benefits of the latter property were first mentioned in Sect. 3.1.1. In addition, the receiver uses a tuned front-end with programmable center frequency and bandwidth to sense the transmitted ELF signal while rejecting out-of-band interference. The receiver front-end was integrated with a data acquisition (DAQ) system to realize a battery-powered portable field unit suitable for on-site deployment in search and rescue missions.

This chapter is organized as follows. Background information and literature review pertinent to ELF field sensing are discussed to put goals in context. The next several sections cover details of receiver design, characterization of the sensor, and realized performance. Finally, the portable receiver is shown.

9.2 Background

Simple air or ferrite-core coils are the B-field detectors of choice at ELF frequencies above a few Hz. While a variety of alternative low-frequency magnetic field detectors exist, they present some disadvantages. Many have limited sensitivity, especially at ELF frequencies (e.g., magnetoresistive devices and Hall-effect sensors). On the other hand, superconducting quantum interference device (SQUID) magnetometers have excellent sensitivity (on the order of 1-10 $fT/\sqrt{Hz}$) at ELF, but are expensive and hard to use since they require cryogenic cooling. Atomic magnetometers, which use an optically pumped vapor of alkali-metal atoms (typically rubidium or cesium), are an appealing alternative since they obtain high sensitivity (comparable to SQUIDs) without the need for cryogenic cooling [1].

J. S. Glickstein and S. Mandal, *Electromechanical Transmitters for ELF/VLF Radio*,
Synthesis Lectures on Engineering, Science, and Technology,
https://doi.org/10.1007/978-3-031-15123-1_9

9.2.1 On Detection and Measurement Sensitivity of Magnetic Fields

This book focuses on magnetic field sensors using air-core coils (i.e., inductive pickup loops). Let us consider here the fundamental physics behind the operation of these sensors.

Consider the case of a circular air-core coil consisting of N turns of wire each with area A_c. A time-varying magnetic flux through the coil induces an emf by Faraday induction,

$$\mathcal{E} = -N\frac{d\Phi}{dt} \tag{9.1}$$

where $\Phi = \oint B \cdot A_c$. The voltage produced across the terminals of the coil by a time-varying magnetic field $B(t) = B_0 \sin(\omega t + \phi)$ which is both uniformly distributed and normal to the plane of the coil is then

$$v_s(t) = N\frac{B_0 A_c \sin(\omega t + \phi)}{t} = B_0 A_c N\omega \cos(\omega t + \phi). \tag{9.2}$$

The coil resistance also contributes a voltage noise equal to the thermal noise $\overline{v_n^2} = 4k_B T R_c$ in series with the induced signal, where T is the absolute temperature and k_B is Boltzmann's constant. The signal-to-noise ratio (SNR) at the coil is

$$SNR = \frac{(B_0 A_c N\omega)^2}{8k_B T R_c}. \tag{9.3}$$

Measurement sensitivity, the amplitude of B_0 when $SNR = 1$ for $\Delta f = 1$ Hz, is

$$S_{min} = \frac{\sqrt{8k_B T R_c}}{A_c N\omega} \propto \frac{1}{\omega A_c^{3/4}} \tag{9.4}$$

which is in units of T/Hz$^{1/2}$ or G/Hz$^{1/2}$ and we have used the fact that R_c is proportional to the coil perimeter, which scales $\propto \sqrt{A_c}$. Some texts refer to this quantity as the "limit of detection" to distinguish it from the "coil sensitivity" of the inductive pickup coil, relating the induced voltage and $\vec{B}$-field magnitude as a quantity in $\frac{V}{T}$. However, this book will maintain the term measurement sensitivity or sensitivity.

9.2.2 Proposed Detector Design

A portable, high-sensitivity 3-axis magnetic field detector is constructed for mobile field use. Even very large (1 m dia., >100 turns) coils can be designed to have a self-resonance frequency above 30 kHz. The bandwidth is adjustable for both relatively broadband detection of modulated signals and restriction to limited channels of interest. The antenna array remains portable for field applications without sacrificing sensitivity.

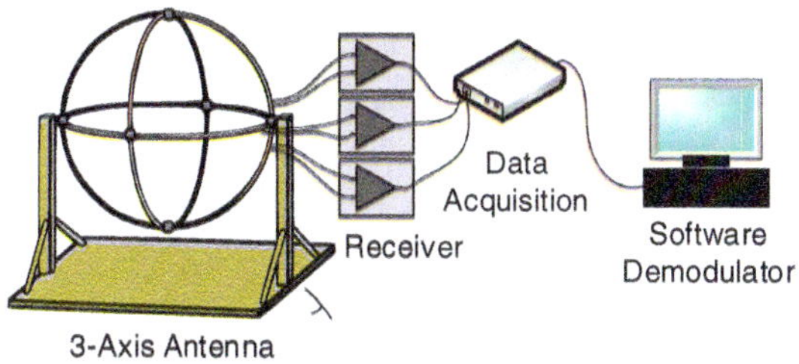

Fig. 9.1 Cartoon depicting the field setup for the ELF receiver

The 3-axis antenna utilizes three identical low-noise preamplifiers, and a data acquisition device captures field data for processing through a software demodulator. Figure 9.1 provides a general depiction of the setup.

9.2.3 Related Work

A system to monitor worldwide thunderstorm activity was presented by Burke et al. in [2]. The authors observe events at ELF frequencies in the 5–45 Hz band using electric and magnetic field antennas. The electric field antenna is a tripod charge-sensing device. Magnetic fields are sensed using twin solenoidal antennas mounted orthogonally. Each antenna consisted of 54,000 turns of copper wire distributed over six laminated ferromagnetic cores to reduce self-capacitance effects. Burke locates the magnetic field sensor in a rural location[1] to reduce interference from power-line harmonics. The field to be detected is on the order of 10 pT. Sensitivity for the magnetic field sensor is reported as $3\frac{V}{nT \cdot Hz}$ at 16 Hz. For the electric field sensor, it is $7\ \frac{V}{m}$.

In other work, a worldwide sensor network is demonstrated by Whitley et al. for sprite studies [3]. The authors focus on the 5–45 Hz band. Orthogonally mounted magnetometers are sourced commercially (MFS-07e, Metronix Geophysics) having sensitivity $0.02\frac{V}{nT \cdot Hz}$ below 32 Hz and $0.64\frac{V}{nT}$ above 32 Hz.

The well-known AWESOME[2] ELF instrument [4] has been extended in recent work by Cohen et al. to a general-purpose ELF/VLF radio receiver for atmospheric weather and VLF radio reception [5]. The receiver operates in the 1–45 kHz band and uses one of two loop antennas, a 1.69 m^2 or 25 m^2 triangle with sensitivity ~ 5 and $\sim 30\ \frac{V}{nT}$, respectively. Both antennas have resistance 1 Ω and inductance 0.5–1 mH. Finally, the same research team has recently extended the broadband "AWESOME" receiver architecture to low-frequency E-field sensing [6].

[1] Chilbolton, England at 51.14° N, 1.44° W.

[2] AWESOME: Atmospheric Weather Electromagnetic System for Observation, Modeling, and Education.

9.3 Receiver Design

9.3.1 Magnetic Field Sensor

Sensitivity is minimized for large-area, low-resistance coils. Consider two circular loop antennas with diameters 2 ft (∼0.6 m) and 3 ft (∼0.9 m), both having N turns of solid 18 AWG copper wire. One fortunate fact is that the skin depth of copper remains twice the radius until 16.5 kHz, well beyond the range of measurement. Simulated resistance, inductance, quality factor, and sensitivity for these two antennas as N varying from 0 to 200 are shown in Figs. 9.2 and 9.3. Quality factor and sensitivity are considered in the range of 120–180 Hz, a plausible communication channel. There is a decisive advantage to large-area detectors. For example, when $N = 100$, despite a 50% increase in coil resistance for $D = 3$ ft over $D = 2$ ft, sensitivity is *improved* by a factor of 50% due to the increase in area. On the other hand, doubling the number of turns only improves this figure by 25%.

There remains a practical consideration, to maintain portability in the field. A 3 ft loop is approximately the maximum size one person can comfortably carry by grasping both sides, and any larger would rapidly push the bound of what can be fit into the trunk of a typical sedan. In addition, the mass of the coil (shown in Fig. 9.4) is just under 5 lbs for $N = 100$ and three are required for the 3-axis receiver. Due to the diminishing return of further increasing the number of turns, this margin is deemed a suitable target.

As an aside, we note that in prior art Cohen et al. demonstrate a triangular antenna with base width 10 m, height 5 m, and area 25 m^2 which has sensitivity 1 fT/Hz$^{1/2}$ at 100 Hz measurement frequency [4]. On the other hand, Fig. 9.3 shows an expected 7 fT/Hz$^{1/2}$ for a proposed circular loop having diameter 0.914 m (3 ft) and area 0.66 m^2. There is certainly

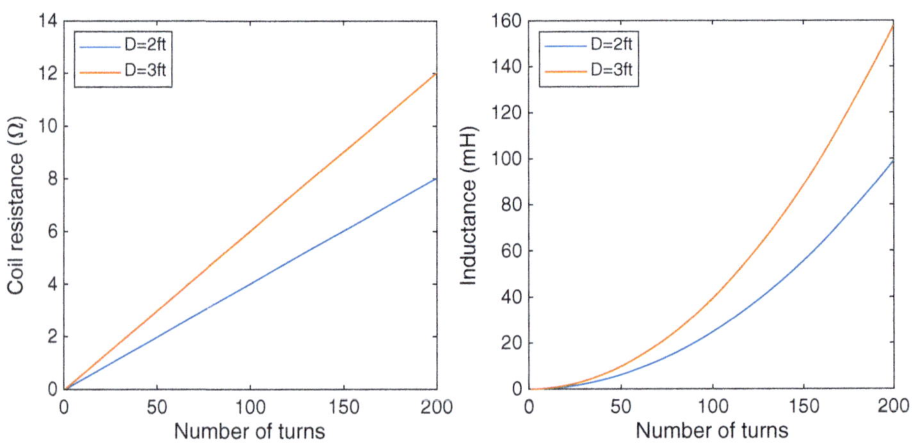

Fig. 9.2 Simulated resistance and inductance of a loop antenna with N turns of 18 AWG copper wire

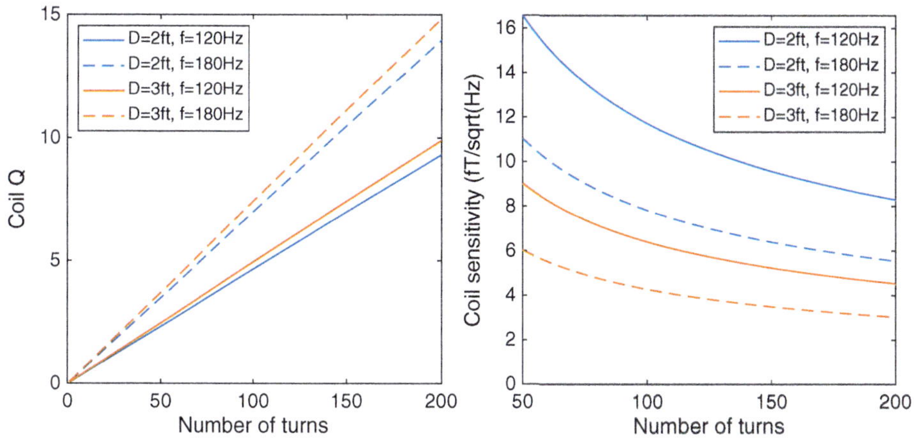

Fig. 9.3 Simulated quality factor and sensitivity of a loop antenna with N turns of 18 AWG copper wire

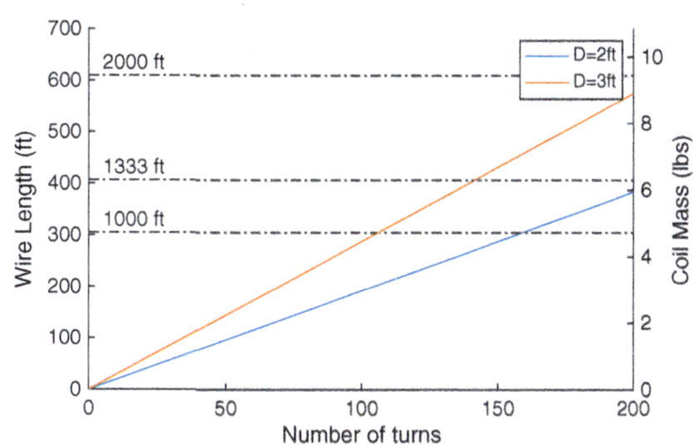

Fig. 9.4 Length of wire in a loop antenna with N turns (left y-axis) and corresponding mass using 18 AWG (right y-axis) in two cases

cause to motivate a large-area antenna despite the loss of portability. Permanent installations may consider this advice in future work.

Three loop antennas are constructed using 1″ portions of Schedule 40 PVC pipe as a structural form. For ease of construction, the circular loop is approximated with an octagon. PVC octagons are assembled by solvent welding of identically sized portions of pipe and 45° plumbing fittings. The structure is allowed to cure for 24 hours before the outer portion of the pipe is removed. Finishing touches (for aesthetic reasons) are completed and the wire is wrapped around the forms and secured. Figure 9.5 shows one completed coil.

Three identical coils were constructed. Impedance measurements were recorded using a network analyzer (Agilent/HP 4395A). Each coil is used as the DUT connected across the two channels, and the analyzer is configured to record cross-channel voltage and phase measurements. We take this approach as the analyzer's impedance test mode is limited to a minimum frequency of 100 kHz.

Fig. 9.5 An octagonal loop antenna with side length $a = 13''$ (0.33 m) and area 0.53 m^2. The diameter of the inscribed circle is 31.5″ (0.8 m)

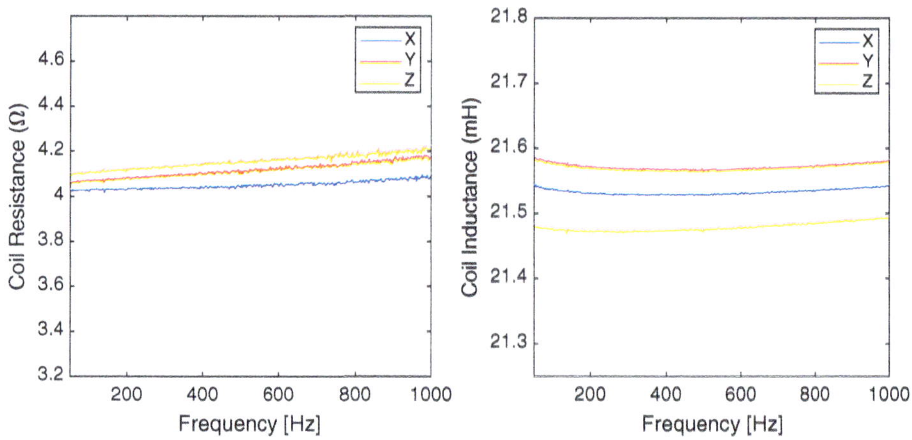

Fig. 9.6 Measured resistance and inductance of loop antenna coils

Measurement results are shown in Fig. 9.6. Parameters are slightly lower than the predicted values because the perimeter of the octagonal loop is slightly lower than for a circular loop with the same diameter. Under these conditions, coil resistance and impedance are in reasonable agreement with the simulated values. Lower inductance is also potentially caused by incomplete coupling between the turns; to reduce this problem, the turns of the coil are pressed into the plastic loop at regular intervals using foam-lined plastic shells[3] held in place by zip ties. Otherwise, the turns of wire tend to stray, which can cause them to snag on something while the antenna is assembled and disassembled on-site. Mitigating the risk of this damage is another plus.

9.3.2 Low-Noise Front-End

A low-noise front-end amplifier has been designed for sub-1 kHz magnetic field detection, where large-area (and thus large Z_s) coils are desirable for high-sensitivity reception. Thus, despite the lower $1/f$ corner frequency in BJT devices, a JFET input stage is preferable to ensure a low input bias current. Readers may refer to classic texts such as Horowitz & Hill's "The Art of Electronics" to find details on noise trade-offs in such devices [7]. The detector is the inductive loop antenna described in the previous sub-section, tuned to resonate within the desired communication channel. The additional voltage gain realized by the resonant network reduces input-referred noise by a factor of Q_c (in rms units) or Q_c^2 (in power units) and thus improves the noise factor, while the limited bandwidth of the tuned circuit rejects out-of-band noise (e.g., power-line harmonics).

The circuit design of the front-end is provided in Fig. 9.7. A fully differential design is used for high common-mode and power supply rejection ratios. The typical noise corner of a low-noise JFET (for example, the NSVJ2394 from ON Semiconductor) lies in the range 100–1000 Hz. This behavior can be observed in noise density measurements of such devices, as displayed in recent work [8, 9]. Each of the input stage JFETs (M1 and M2 in Fig. 9.7) was implemented as N parallel devices to reduce the $\frac{1}{f}$ noise PSD by a factor of $\frac{1}{\sqrt{N}}$. The required number varies based on the device chosen and the required sensitivity of measurement. We take 16–32 as the nominal value of N for a sensor of size described in Sect. 9.3.1.

The sensor network at the input to the preamplifier is shown in Fig. 9.8. Adaptive tuning of the resonant network uses an adjustable tuning capacitance. We also apply noiseless bandwidth adjustment (akin to Q-damping) by adjusting the value of the capacitors in feedback with the preamplifier [10]. The feedback capacitance in this case is made adjustable with a switched network. At the particularly low frequencies of ELF communication and in such a low voltage front-end, bidirectional CMOS switches are ideally suited for their ease of implementation.

The front-end is followed by a post-amplifier which converts the differential output to a single-ended signal, applies additional gain, and optionally provides notch filtering at two frequencies. The layout is fairly standard and consists of several CMOS op-amp stages; it is not shown here. When in use, the circuit will be operated from batteries. Two 3.7 V Li-ion cell batteries serve as a dual-rail supply.

[3] Made from pieces of 3/4" PVC pipe which are cut in half, a perfect fit inside a 1" PVC pipe.

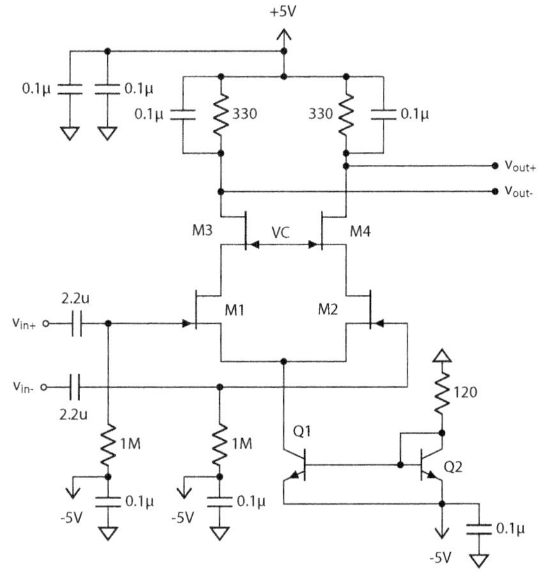

Fig. 9.7 Analog front-end of the ELF receiver

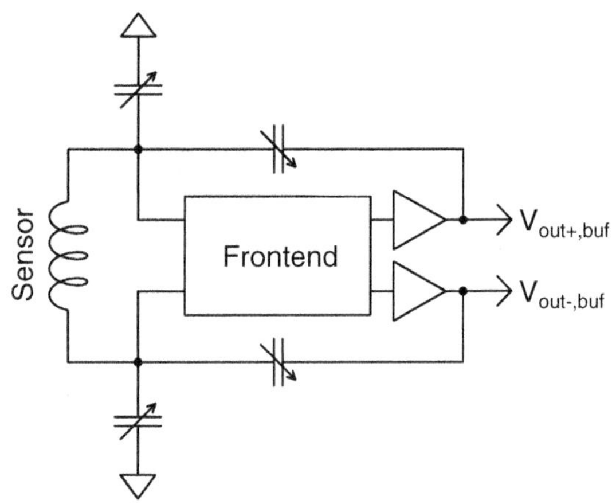

Fig. 9.8 B-field sensor network at input to analog front-end; both tuning and feedback capacitors (C_T and C_f, respectively) are shown

The receiver utilizes a DAQ (USB-6259, National Instruments) for data acquisition. This DAQ has 16-bit ADC resolution and achieves a maximum aggregate multichannel sample rate of 1 MS/s. Post-amplifier gain is adjusted based on permissible output swing and anticipated peak field amplitude. Note that the amplifier design includes conversion to single-ended output, based on intended use with an embedded ADC in future work.

Using simulated values for an antenna consisting of 100 turns of 18 AWG copper wire (40 mH, 6 Ω), the AC transfer function characteristics of the detector are simulated in LTSpice (a widely used free circuit simulation program). Figures 9.9 and 9.10 show the gain and noise factor, respectively, for various values of tuning capacitance. Note that the low-

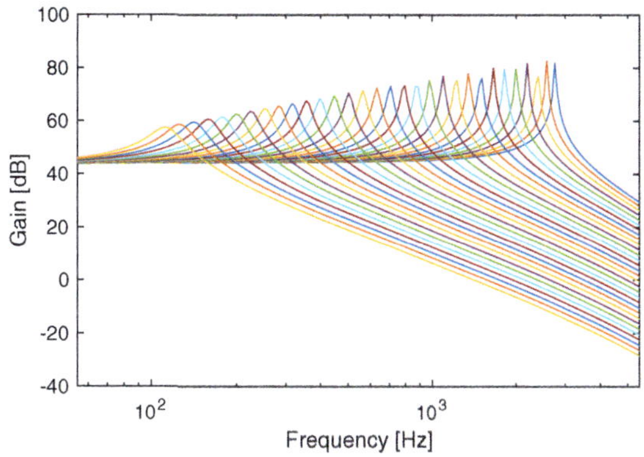

Fig. 9.9 Simulated AC transfer function of receiver and sensor for varied tuning capacitance $C_T = 0.1–100\ \mu F$

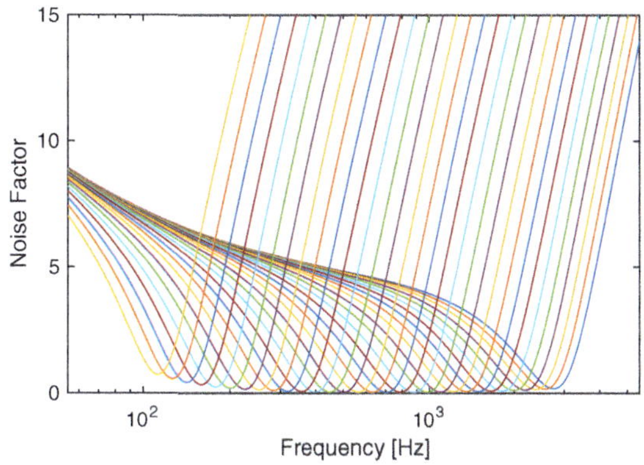

Fig. 9.10 Simulated noise factor of receiver and sensor for varied tuning capacitance $C_T = 0.1–100\ \mu F$

frequency cut-in is too low to be observed; this is intentional to avoid reducing measurement sensitivity. In future work, it should be possible to adjust for reduced saturation by 60 Hz interference (see future work, Sect. 11.2.2).

Bandwidth adjustment by active Q damping allows receiver to operate over a broad channel as needed. Simulated bandwidth adjustment by Q damping is shown in Fig. 9.11 under the same operating conditions as previously described. The tuning capacitance is 47 μF in this case.

9.3.3 Measured Characteristics

Device variations lead to some differences in gain between multiple discrete circuits. The gain for each of the three amplifiers used for the 3-axis receiver must be comparable, and they are made adjustable using a variable gain CMOS operational amplifier circuit after conversion to single-ended output. Figure 9.12 shows the measured gain for three circuits after they are adjusted to be approximately equivalent.

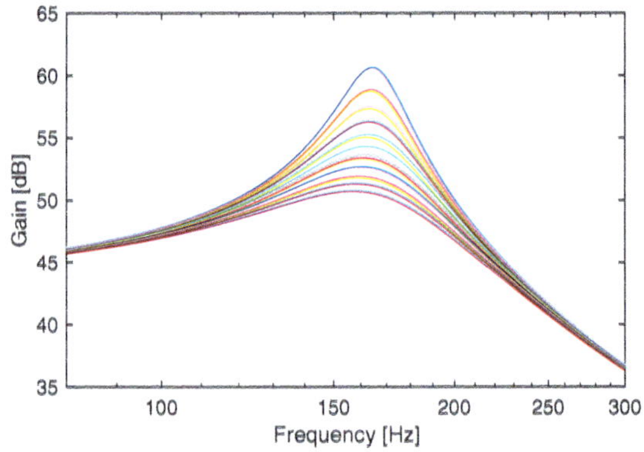

Fig. 9.11 Sweep of gyrated feedback capacitance $C_f = 0.1$–20 nF for typical detector tuned to 165 Hz

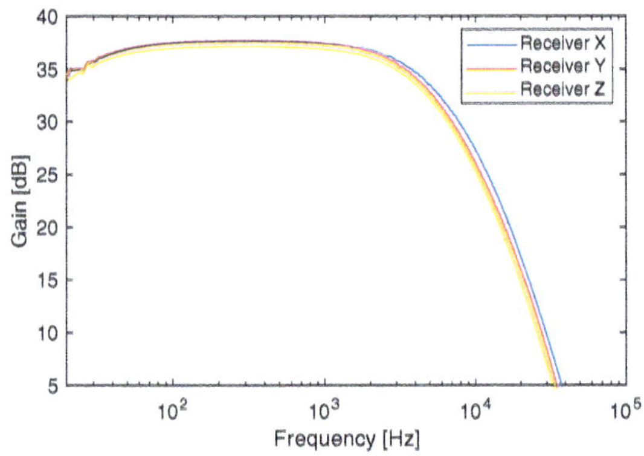

Fig. 9.12 Measured amplifier gain for three assembled circuits, each powered by batteries

Input-referred noise of the circuit is computed by the standard tactic of measuring output-referred noise using the spectrum analyzer and dividing by the measured gain. Data are shown in Fig. 9.13. The circuit is again battery-powered and installed in a shielded enclosure. Data are collected using two measurement ranges to ensure sufficient points are collected over the full frequency interval. Integrating factor bandwidth is reduced for low-frequency measurement. This approach results in a discontinuity of sampling frequency step size at 10^3 Hz and an apparent drop in noise variance, which is purely visual.

Tuning function of the receiver was verified by measuring ambient noise over a period of long ($T = 100$ s) acquisition time. The white noise of the spectra shows distinguishable peaks at the center frequency of tuning. Results are presented in Fig. 9.14; note that here the data are windowed to suppress low-frequency content with flatband response as this portion of the data prevents the tuned response at multiple measurement points from displaying clearly in the figure. In the actual measurement, low-frequency content has approximately constant and equal amplitude in all cases (as shown in the simulation).

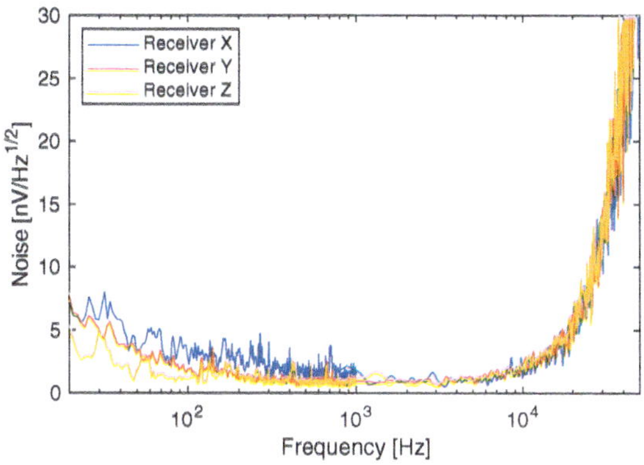

Fig. 9.13 Measured input-referred noise of three assembled circuits

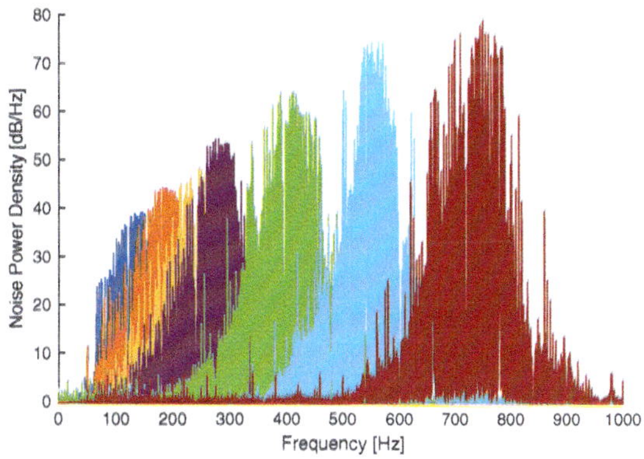

Fig. 9.14 Measured white noise using the receiver tuned to various frequencies by adjusting the tuning capacitance

9.3.4 3-Axis Antenna Calibration

Field measurements were calibrated by generating single frequency tones at step intervals across the application channel; for this purpose, the AMEBA2 does the job as a field generation source. The detector and a known reference source are both positioned at the same fixed distance away from the transmitter.[4] Data was collected for several seconds to identify peak field amplitude.

The reference channel measurement is made using the same sensor and amplifier as the prototype receiver, as detailed in Sect. 5.2.2. The amplifier is characterized using the same network analyzer as is used to collect measurements of the new receiver's performance. Transfer function and noise characteristics of the reference channel amplifier are produced in Fig. 9.15; we find the gain and noise performance match the manufacturer's specifications.

Utilizing the reference channel, the $\vec{B}$-field amplitude at the center of the coil-under-test is measured. We take measurements at several query frequencies in the range where we know the calibrated reference sensor has a flat transfer function. We find the measurement results,

[4] 8 ft (2.44 m).

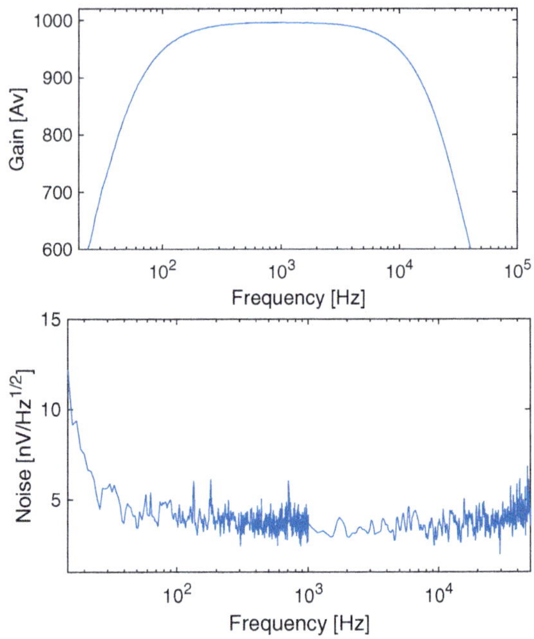

Fig. 9.15 Gain and noise characterization of amplifier used in reference channel

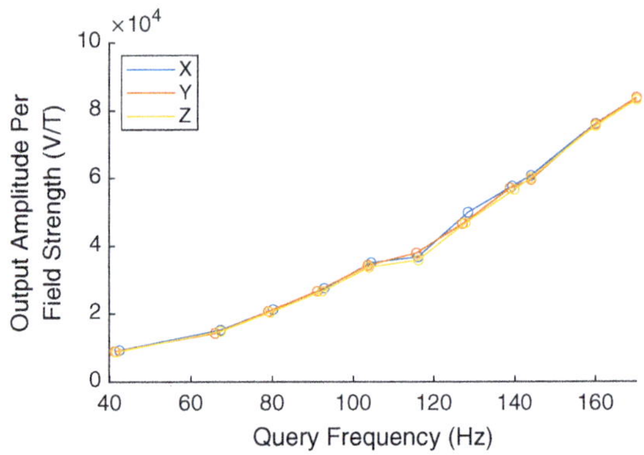

Fig. 9.16 Receiver output voltage per unit $\vec{B}$-field collected for all three antennas

which are summarized in Fig. 9.16 for all three antennas, to be in good agreement with the theoretical value computed for AMEBA2 using the same approach as for the prototype.[5]

Using the recorded amplifier output and computed field strength measurements, the gain (amplitude response) of each pair of antenna and amplifier board are determined. This gain relates to the output amplitude of the receiver per unit field strength and is displayed in Fig. 9.17. Measurements are recorded after the application of digital filtering in software to remove harmonics at $N \times 60$ Hz, otherwise the amplitude of the single test tone cannot be

[5] Section 5.3.1, specifically Fig. 5.11.

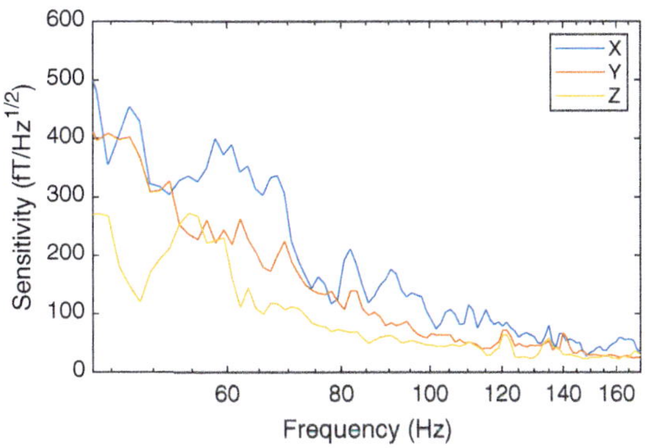

Fig. 9.17 Measurement sensitivity for the ELF receiver

readily determined. We can observe an interesting result near 120 Hz where the measured field is skewed compared with the rest of the data. This is likely due to the challenge of removing 120 Hz interference without disturbing the desired signal at the probe frequency (115 Hz).

Measurement sensitivity is computed from realized sensor gain, using the sensor noise of each antenna as characterized by a dynamic signal analyzer. The results are shown in Fig. 9.17. The sensitivity above 130 Hz is below 40 fT/Hz$^{1/2}$.

9.4 Portable Field Unit

Figure 9.18 shows the portable field unit deployed for testing purposes. It integrates the 3-axis receive coil (assembled inside plastic tubing to ensure robustness to transport and weather conditions) with the preamplifiers (assembled inside a shielded metal box to minimize external noise pickup) and the DAQ unit (connected to a laptop over USB). The system is designed for ease of use and rapid assembly in the field.

9.5 Performance and Concluding Remarks

Two limitations exist in the determination of effective transmit distance. The first is the conditions of the channel and the resulting E_b/N_0 at the receiver. Techniques have been explored to increase E_b using additional transmit cells, and to reduce N_0 by oversampling. N_0 may also be reduced by selecting a favorable position for the receiver, though this is not always possible. The second limitation is the noise floor of the receiver, which is dependent on the noise floor of the antenna itself and the first stage of the preamplifier.

The new ELF receiver presented in this chapter achieves a measurement sensitivity of 50 fT/Hz$^{1/2}$ above a 100 Hz measurement frequency. Above 130 Hz, the sensitivity improves below 40 fT/Hz$^{1/2}$. A crucial metric is to observe that this is lower than the noise floor of

Fig. 9.18 Portable ELF receiver deployed in residential driveway for evaluation

ambient EM field levels in all locations observed in our site surveys (described in Sect. 4.2). To summarize, we observed a noise floor for the ELF channel which lies anywhere between -230 and -170 dBT/Hz (corresponding to 3 pT/Hz$^{1/2}$ and 3 nT/Hz$^{1/2}$, respectively). The large variance is part of the interesting challenges of ELF communication.

After removing man-made sources (e.g., power-line harmonics), geomagnetic noise (known as "micropulsations") due to time-varying interactions between the ionosphere and the solar wind is the dominant ELF noise mechanism below ~10 Hz [11]. The power spectrum of this mechanism decreases as approximately $1/f^{1.3}$. In addition, peaks in the power spectrum corresponding to the Schumann resonances are often visible (e.g., the first resonance at ~7.8 Hz). At higher frequencies, the ELF/VLF noise floor is dominated by broadband noise generated by lightning (e.g., from distant thunderstorms). Since lightning is much more frequent in summer than in winter, the ELF/VLF noise floor is also seasonally varying.

Even after taking variance into account, we can safely conclude that the ambient noise will dominate N_0 in the established data link. Compared to the prototype receiver of Sect. 5.3, the new receiver has a measurement sensitivity that is a factor of $8\times$ lower.

We also consider a comparison to the AWESOME instrument [4], which achieves 40 fT/Hz$^{1/2}$ measurement sensitivity at 100 Hz, dropping to 15 fT/Hz$^{1/2}$ at 1 kHz. The AWESOME instrument uses a 1.69 m^2 area triangular antenna, a factor of ~$3.2\times$ larger than the area of our loop antenna. Compared to the AWESOME instrument, our ELF antenna is more portable while still remaining more sensitive than the noise floor of the environment.

References

1. Kominis, I. K., Kornack, T. W., Allred, J. C., & Romalis, M. V. (2003). A subfemtotesla multichannel atomic magnetometer. *Nature, 422*(6932), 596–599.
2. Burke, C. P., & Jones, D. L. (1995). Global radiolocation in the lower ELF frequency band. *Journal of Geophysical Research: Atmospheres, 100*(D12), 26263–26271.
3. Whitley, T., Füllekrug, M., Rycroft, M., Bennett, A., Wyatt, F., Elliott, D., et al. (2011). Worldwide extremely low frequency magnetic field sensor network for sprite studies. *Radio Science, 46*(04), 1–9.
4. Cohen, M. B., Inan, U. S., & Paschal, E. W. (2009). Sensitive broadband ELF/VLF radio reception with the AWESOME instrument. *IEEE Transactions on Geoscience and Remote Sensing, 48*(1), 3–17.
5. Cohen, M. B., Said, R. K., Paschal, E. W., McCormick, J. C., Gross, N. C., Thompson, L., Higginson-Rollins, M., Inan, U. S., & Chang, J. (2018). Broadband longwave radio remote sensing instrumentation. *Review of Scientific Instruments,89*(9), 094501.
6. Gurses, B. V., Whitmore, K. T., & Cohen, M. B. (2021). Ultra-sensitive broadband "AWESOME" electric field receiver for nanovolt low-frequency signals. *Review of Scientific Instruments,92*(2), 024704.
7. Horowitz, P., & Hill, W. (2015). *The art of electronics* (3rd ed.). Cambridge: Cambridge University Press.
8. Danyuk, D. (2013). Measurements Rate SMT Low-Voltage n-JFETs Under Consistent Conditions. Electronic Design. (Online). https://www.electronicdesign.com/power-management/article/21796240/measurements-rate-smt-lowvoltage-njfets-under-consistent-conditions.
9. Danyuk, D. (2021). Measurements Rate New SMT Low-Voltage n-JFETs Under Consistent Conditions - An Update Using Modern JFETs. audioXpress. (Online). https://audioxpress.com/article/measurements-rate-new-smt-low-voltage-jfets-under-consistent-conditions-an-update-using-modern-jfets.
10. Mandal, S., Utsuzawa, S., Cory, D. G., Hürlimann, M., Poitzsch, M., & Song, Y. Q. (2014). An ultra-broadband low-frequency magnetic resonance system. *Journal of Magnetic Resonance, 242,* 113–125.
11. Romani, G. L., Williamson, S. J., & Kaufman, L. (1982). Biomagnetic instrumentation. *Review of Scientific Instruments, 53*(12), 1815–1845.

A Complete ELF Communication System 10

10.1 Field Setup: Transmitter

A self-contained cave radio transmitter was constructed to contain the electromechanical transmitter of Chap. 7 and control electronics of Sect. 8.1.1. The device is introduced here as we have finally covered all pertinent background information on sub-components within. An auxiliary air supply tank converted to a backpack form factor serves as an external compressed air "battery".

The portable transmitter of Fig. 10.1 is capable of operating with the lid open or closed. The device's electrical power supply is in the form of three sealed 12 V lead acid batteries located in the bottom of the enclosure. One 12 V battery supplies the control system (and optionally a communication terminal connected externally). Two 12 V batteries in series provide the 24 V supply required by the BLDC motor. This approach is motivated by convenience to maintain balance between the series batteries supplying the motor and avoid the need for a step-down converter to supply the control system from 24 V.

Air is supplied to the transmitter from an external tank. Filtration system components are attached to the lid of the transmitter apparatus downstream from a final pressure regulator to ensure maximal air quality is attained. These components are oriented correctly for operation when the lid of the radio device is opened as shown in Fig. 10.2.

A 1999 Jeep Wrangler was commissioned as the AMEBA2 emergency cave radio communication system deployment truck. To demonstrate portability of the experimental setup, Fig. 10.3 shows the complete transmitter and receiver in the cargo area of the Jeep.

J. S. Glickstein and S. Mandal, *Electromechanical Transmitters for ELF/VLF Radio*,
Synthesis Lectures on Engineering, Science, and Technology,
https://doi.org/10.1007/978-3-031-15123-1_10

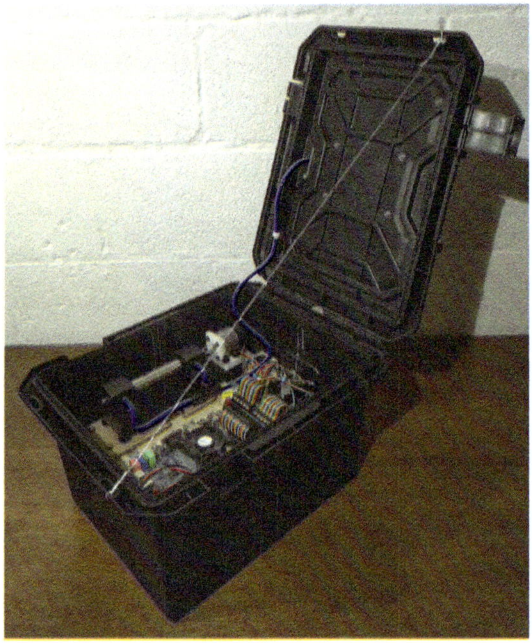

Fig. 10.1 Portable cave radio transmitter opened to show internal components

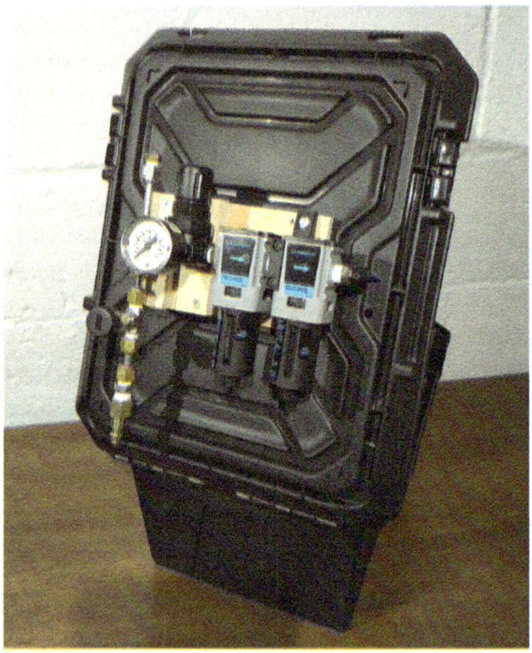

Fig. 10.2 Air pressure regulator, particulate filter, and drier are attached external to the transmitter after a series flow regulating valve

Fig. 10.3 Portable transmitter and receiver packed into mobile AMEBA truck ready for transportation to use site and deployment

10.2 Results of Line of Sight Testing

10.2.1 Single Tone Transmission

We began testing by transmitting a single tone with 10 m separation between the transmitter and receiver, already over 2× greater distance than was achieved by the prototype AMEBA. Figure 10.4 shows the spectral spread of such a measurement over a 10 s acquisition window in time domain. To demonstrate effectiveness of the controller action we show the spectral spread of this tone transmitted in open loop and closed loop control regimes. Notice that the open loop control shows spectral spread, while the closed loop controller response produces a tone having tight spectral spread and slight undershoot of the target. This is an indicator that more aggressive controller action (i.e., increased proportional gain) might be required.

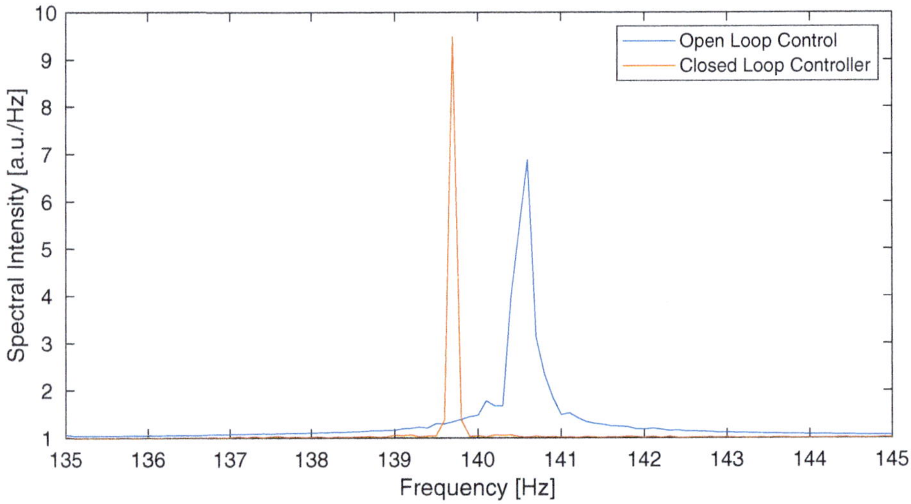

Fig. 10.4 Power spectrum of single tone transmission at 140 Hz by AMEBA2 showing spectral spread with and without controller action

10.2.2 Established CP-FSK Data Link

Despite the 2.63% reduction in power consumption for CF-FSK modulation compared with CP-FSK (see Sect. 8.4.4), we switch to CP-FSK for the established data link in this section to effectively demonstrate the capabilities of the transmitter. The receiver decoding scheme for CF-FSK demodulation becomes too complex for the scope of this project and is reserved for suggested future work.

We continue to work with the message,

ALL THINGS EXIST AS THEY ARE PERCEIVED...

but the packets are reduced to contain four symbols with a symbol period $T_s = 2$ s in order to allow time-frequency analysis by visual inspection. We will proceed to establish a data link and analyze recorded content of time domain data using spectrograms. The intent of our analysis is now strictly to show that the transmitted waveform can be recovered by the receiver.

Packets are converted to four symbols by placing a single symbol in each word, to maintain the datatype organization of Sect. 8.3.1 for the message. Parity symbols precede the message contents in each packet and are given a shorter symbol period. We allow the transmitter to return to the carrier frequency between packet transmissions, so the parity symbol operates as a start signal in addition to conveyance of data integrity signaling for error checking. A sequence of four packets modulated by this approach are shown in Fig. 10.5.

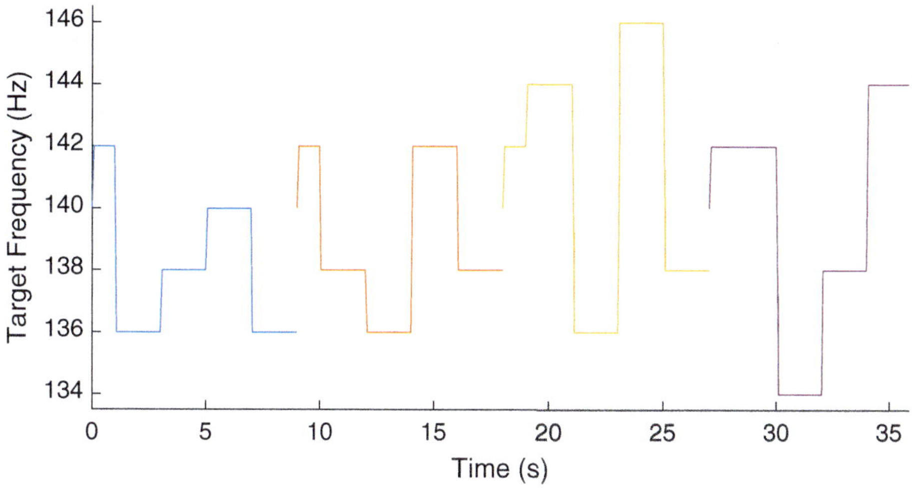

Fig. 10.5 Modulated signal of 4 packets each containing 4 symbols. Packet content are preceded by the parity symbol

We perform a short site survey and select a carrier frequency of 140 Hz, identifying the 134–146 Hz channel to be free of spurious noise.

On the receiver, we record data over 120 s at a 20 kHz sample rate. Oversampling reduces uncertainty in time domain analysis and improves the results of sinusoidal curve fitting. A spectrogram of the data is computed using a 0.5 s window size so the frequency resolution is equal to the space between levels Δf in the transmission. Since the parity symbol contains only one bit of information and always follows a tone at the carrier frequency, it is easy to distinguish its maximum likelihood value in the receive sequence. While the somewhat long duration between packets is inefficient in terms of data rate, it is convenient for identifying the beginning of a packet.

Plotting the peak power spectral density at each point in time is useful to show the ease with which an automated receiver could identify the start of packets. Figures 10.6 and 10.7 show an example. We also see how some of the points in each packet deviate from the correct value, particularly at the symbol boundaries.

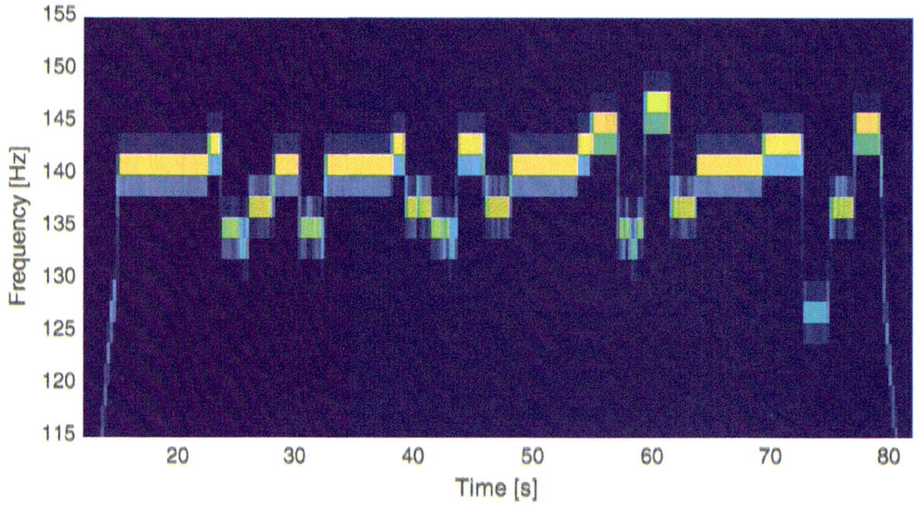

Fig. 10.6 Spectrogram of received data over a 10 m channel in open air. Four packets are visually distinguishable from the background

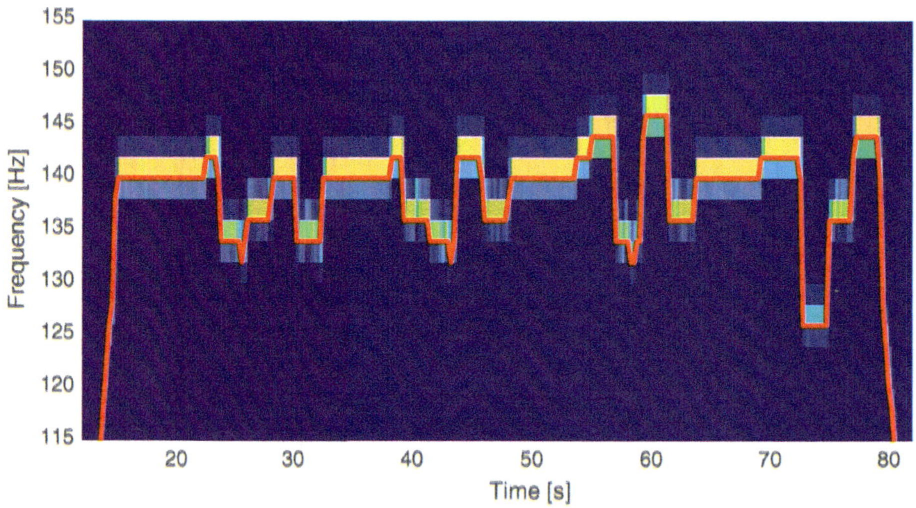

Fig. 10.7 Peak power trace plotted over the spectrogram of Fig. 10.6

Future work should explore automated symbol sequence recovery by identification of these boundaries followed by a time domain fit of the symbol data. Time domain data corresponding to the spectrogram of Fig. 10.7 is easily usable to extract the symbol sequence provided some care is taken to identify symbol boundaries. Notch filtering at high intensity noise sources followed by a band reject filter to remove out of band signals is useful here.

Due to the closeness of noise sources to the signal, the band reject filter has limited usefulness and is advised with caution (it is easy to distort the desired signal and its phase). Power-line harmonics can often be removed by applying a comb filter (series of notch filters at multiples of 50/60 Hz) though this is sometimes made more difficult by localized variance in harmonic power and spread. In future work, routes to a generalized set of guidelines for detection and mitigation of these harmonics should be explored. The authors suggest use of the power spectrum over a complete message to identify noise peaks and their bandwidth.

10.2.3 Maximum Link Distance

Maximum effective link distance was experimentally verified by measuring the field drop-off over distance in a residential suburb of New Albany, Ohio, USA. The magnetic field amplitude was measured at controlled distances away from the transmitter during an experiment at approximately 9 PM at night in January with no snow on the ground. We conduct the measurement on the sidewalk at a minimum of 10 m from the nearest power line.

Figure 10.8 shows line of sight data acquisition in progress. The receiver (at left) is oriented with y-axis parallel to the direction of the sidewalk. Street lights in the neighborhood are gas mantle lamps and power distribution lines are buried on property boundaries behind homes, making this an ideal location for a field test. Man-made interference at power-line harmonics is somewhat lower in this location relative to other suburban locations.

Figure 10.9 shows the experimental data and a fit to $|\vec{B}| \propto \frac{k}{r^3}$. The carrier frequency 130 Hz is selected by site survey to first verify the available frequencies in the channel. On

Fig. 10.8 Receiver (at left) and transmitter (at right) setup across snowy sidewalk of suburban central Ohio, USA. Picture taken January 2022

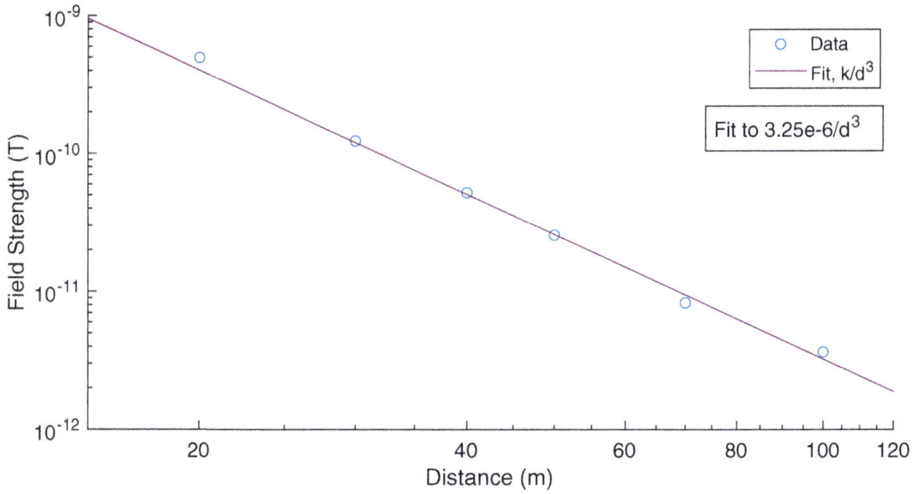

Fig. 10.9 Field verification of rate-of-decay trend in magnetic field generated by electromechanical transmitter at 130 Hz

the particular day of this experiment, noise at 140 Hz motivated[1] the use of an alternative carrier.

At 100 m from the transmitter, a 3.67 pT field is successfully recorded. At the particular time of measurement, the flatband noise floor of the ELF channel was approximately 2 pT. Based on the measured trend we expect the data link range in this environment is 117 m. Since the noise floor is known to vary seasonally and by time of day, it is important to conduct a site survey prior to all experiments and we record the ambient EM field PSD to (i) identify a suitable channel with maximally flat in-band noise distribution, and (ii) estimate the functional limit to communication based on ambient noise conditions.

At the noise floor of the ELF field receiver (40 fT/Hz$^{1/2}$), the B-field of the electromechanical transmitter is projected to have an effective range of ~475 m in open air.[2] Based on our measurements of typical background noise, we know that ambient noise of the environment will set a higher noise floor in practice, which is 100 fT/Hz$^{1/2}$ in the best case (corresponding maximum link distance estimation is 350 m). However, realistically the link distance may be well below this value, dependent on daily conditions.

[1] Man-made interference at arbitrary frequencies in the ELF channel is not uncommon and also difficult to predict due to the large assortment of possible devices inside residential homes.

[2] Maximum link distance based on the point at which $E_b = N_0$ over a 1 Hz bandwidth (SNR of 1).

10.3 Cave Deployment

This section is written from the perspective of a *simulated* emergency situation to demonstrate the field efficacy of AMEBA2. Within this section the term "emergency" is used to present a theoretical context only.

On a snowy winter day in the first month of 2022, one of the authors (J. Glickstein) embarks on a journey to West Liberty, Ohio, USA to deliver an emergency radio communication system to the Ohio Caverns, a publicly accessible set of caves that advertises itself as "Ohio's Largest and Most Beautiful Cavern". The radio is transported through harsh weather conditions as shown in Fig. 10.10, with several inches of snow already on the ground. Outside, the temperature is a high of −1° C (30° F) and low of −5° C (23° F).

Arrival at the gates of Ohio Caverns is shown in Fig. 10.11. At the caverns, we meet our host Tim, who introduces us to one of his exploration experts, Ben. Ben directs us into the cave and shows us where to set up the transmitter to send an emergency signal to the surface. Inside the cave, we find ourselves located approximately 15 m below ground. The nearby exit passageway is barricaded by ice buildup due to inclement weather conditions.

Fig. 10.10 Field work requires one of the authors to drive through poor weather conditions to deliver emergency cave radio to remote location

Fig. 10.11 Brave adventurer (one of the authors) standing next to field deployment vehicle at the gates to Ohio Caverns in West Liberty, Ohio, USA

The ELF receiver is deployed on the surface above the cave. Figure 10.12 shows the assembled receiver in the middle of the snow. We identify the surface position approximately above the in-cave location using a man-made landmark. Our on-site guide tells us the landmark was positioned using reference maps drawn during early cave exploration. We perform a site survey and verify the local 134–146 Hz channel has minimal and flat in-band noise, enabling use of a 140 Hz carrier (the same as in previous line of sight testing).

Returning to the cave, the transmitter is set up on a table where space permits. Inside the cave, the temperature is a constant 12.2° C (54° F) and water droplets occasionally fall from the moist limestone surfaces overhead. The rugged, waterproof enclosure protects the transmitter from the elements. Figure 10.13 shows the transmitter in action.

Fig. 10.12 Receiver setup on the surface above the cave

The experiment begins with single tone transmission over 40 s at two frequencies to enable easy identification of the test signal. In the spectrogram of Fig. 10.14 these tones are easily visible and validate the transmitter signal can be detected through the rock at the surface. There is no line of sight and no alternative form of communication between the cave and the surface. Data acquisition on the receiver is performed asynchronously using a continuous acquisition over 100 s intervals. Each acquisition is time stamped. In the cave, we record the time of transmission when operating the transmitter. After sending a message, we return to the surface and post-process data from the receiver to identify the successful reception of a cave-to-surface signal.

Fig. 10.13 The first author uses the transmitter in the cave to send a message to the surface

We repeat the same process for a complete message, in this case a series of 8 characters sent over four packets which are successfully recovered on the surface. A spectrogram of the data is shown in Fig. 10.15, recorded from a cave-to-surface transmission through an

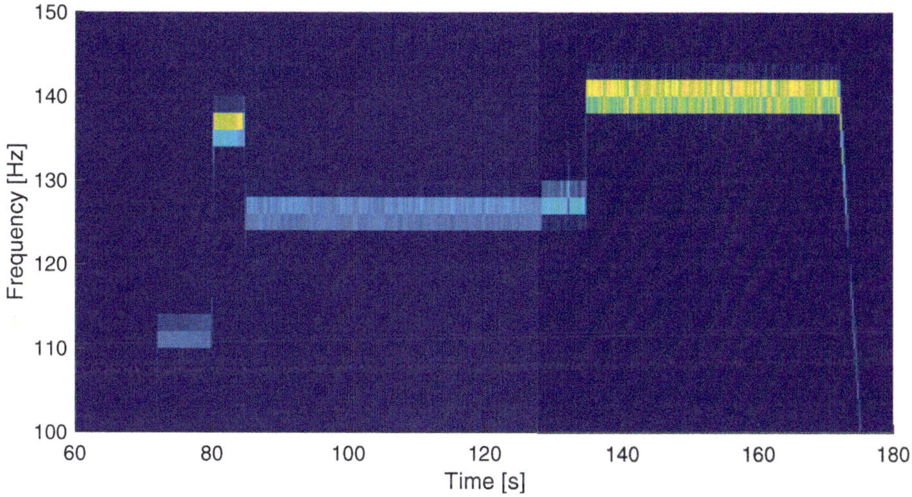

Fig. 10.14 Spectrogram for time-domain acquisition of single tone cave-to-surface transmission at 125 and 140 Hz to verify signal from ELF transmitter radiates through the rock

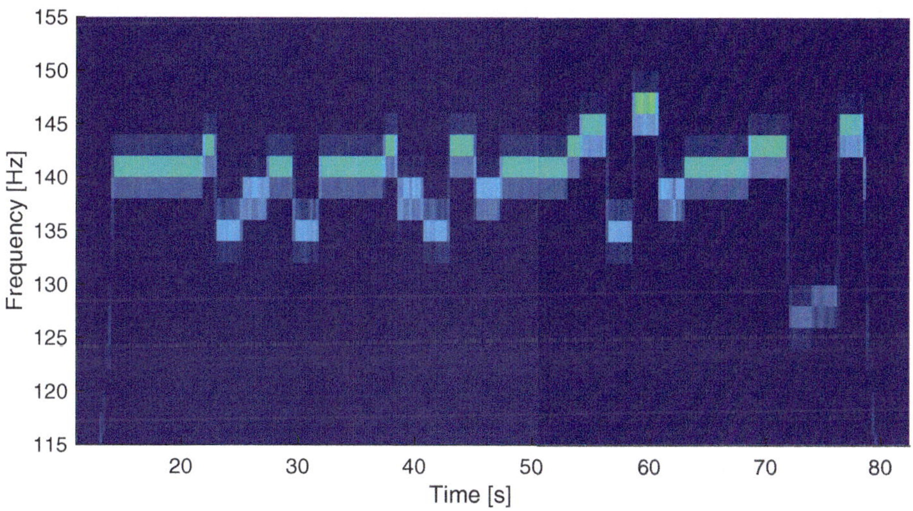

Fig. 10.15 Spectrogram showing first 16 symbols of encoded test message as received during cave-to-surface transmission at carrier frequency 140 Hz

estimated 15 m of rock. Our guide tells us there are three layers between the cave and surface: limestone, shale, and frozen soil. We depart the Ohio Caverns feeling gratified by the successful test results.

Conclusions and Future Work 11

11.1 Conclusions

11.1.1 Introduction to the AMEBA Concept

Portable and low-power ELF transmitters are necessary for two-way wireless communications with submarines, mines, and tunnels (e.g., during emergencies when wired links fail). Even a relatively slow link (tens-hundreds of bits/sec) capable of sending short text messages within a few seconds would be sufficient in these situations. Other promising applications of miniaturized ELF transmitters include geophysics (e.g., subsurface imaging for locating geological conductors, such as ore deposits) [1] and atmospheric science (e.g., ionosphere research) [2, 3].

In the absence of effective RF communication systems in the undersea and underground environments, a wide assortment of communication modalities adapted to the challenges of each environment have developed to meet demands by military, industrial, and recreational applications. While these modalities have proven to be somewhat adequate, the need for a more robust solution persists.

11.1.2 Implementation of an Open-Loop AMEBA Prototype

The design of a prototype miniaturized power-efficient ELF transmitter based on the mechanical rotation of permanently-polarized dipoles was shown. The latter can use permanent magnets (ferromagnets, suitable for ELF) or electrets (ferroelectrics, suitable for VLF) [4]. The moment of inertia of the rotating dipoles must be minimized to reduce peak electrical power and increase the data rate. The use of a distributed array of $N > 1$ sub-elements minimizes overall inertia, while long and thin cylinders are optimal geometries for minimizing

J. S. Glickstein and S. Mandal, *Electromechanical Transmitters for ELF/VLF Radio*, Synthesis Lectures on Engineering, Science, and Technology,
https://doi.org/10.1007/978-3-031-15123-1_11

the inertia of each sub-element. However, such elongated shapes are difficult to manufacture in practice. The high-speed rotation of the magnetized cylinders is also accompanied by formidable mechanical design challenges (balancing, dynamic stability, etc.), which were addressed in depth.

An ELF transmitter constructed using a mechanically-rotating magnetic dipole was evaluated to demonstrate its use for non-line-of-sight wireless communications in conductive media. Data was successfully transmitted through walls and floors, in particular from inside a laboratory to a hallway on a different floor of the building. Design challenges were considered for the receiver, including the impact of powerline interference on the effectiveness of long-range data transmission.

11.1.3 Design of a Closed-Loop AMEBA Transmitter

Design constraints for the transmitter and the scope of their effects were put in context. The mechanical transmitter system was constrained in the mechanical domain and analyzed to describe tradeoffs in design. We observed that the electrical domain is equally constrained and established relations between design guidelines described for each. In Sect. 6.3 all components of the transmitter were reduced to an equivalent torque including sources of power input and loss from the electrical domain.

Following this analysis, Sect. 6.3.4 presented a model for the optimization of magnet geometry to achieve peak field generation efficiency in the AMEBA2 unit cell. Using the model the optimization parameters for the system were identified. Given a need for minimum inside diameter and maximum length, we found that the values are fixed by the limits of manufacturing capabilities. The performance impact of the optimum magnet and its scope are now well defined. In addition, the performance impact of communication parameters and the channel were considered in Sect. 4.2.

11.1.4 Implementation of the Closed-Loop Transmitter

Practical considerations of the mechanical design for AMEBA2 were discussed. The rotating assembly was assembled, balanced, and adjusted per limitations of manufacturing, machining, and real world needs. A power amplifier to drive the BLDC motor was designed using a discrete GaN FET based three phase bridge. The final assembly required hand adjustment and shimming to achieve suitable alignment and balancing.

11.1.5 Transmitter Control System

The BLDC motor was modeled using a block diagram and simulations were used to describe the function of the device. Experimental data on the performance characteristics and step response of the plant was collected in open loop. The experimental data showed that the system can be approximated as a FOPDT model, thus confirming that the design achieves our goal of modulation energy dominating the power consumption of the device. A closed loop controller was implemented in FPGA hardware using a discrete logic implementation of a classical PID controller. After tuning, the controller achieves good stability with a phase margin of 33.9° using a modest control regime. Power consumption using CF-FSK modulation was observed to be reduced by a factor of 2.63% compared with CP-FSK, confirming that the scheme reduces energy use as designed.

The AMEBA2 achieved a factor of 4.9× greater operational power efficiency than the prototype. In addition, AMEBA2 is 7.6× more power efficient than an equivalently sized electrical antenna. In terms of energy use, the improvement is more dramatic due to the closed loop control system. In the field tests, the energy used to transmit an identical message was a factor of 9.8× lower than for the prototype due to the reduced symbol rate. A reduction in symbol rate down to $R_s = 50$ ms is possible using conservative settings for the controller, at which point the symbol rate is limited by the minimum which maintains orthogonality of the set. Using Δf of 2 Hz, the symbol rate could be reduced to 0.5 s, thus reducing energy consumption per message by another factor of 4× (and allowing an SMS-length text message to be sent in under 3 min).

11.1.6 Design and Implementation of an ELF Receiver

A portable, state of the art high-sensitivity ELF receiver was designed and constructed for field test demonstration of the AMEBA2 transmitter. The receiver has 3-axis vector measurement capability and uses three 0.8 m diameter loop antennas with 100 turns of 18 AWG copper wire used for each loop. Sensitivity above 130 Hz is < 40 fT/Hz$^{1/2}$, which is lower than the noise floor of the environment. The receiver can be readily flat-packed for portability.

11.1.7 A Complete ELF Communication System

The mechanical transmitter was integrated into a portable, fully enclosed unit ready for field test deployment. A communication link was established in open air in a residential suburb under line of sight conditions and data was sent across the established link with a separation of 10 m. The maximum link distance for this setup was then measured; the signal from the transmitter was detected at distances up to 100 m, at which point the signal-to-noise ratio (SNR) was only 1.8.

The transmitter was then taken to a cavern in rural Ohio, USA, where we demonstrated successful cave-to-surface transmission of a message through 15 m of rock and frozen soil. Both transmitter and receiver held up to abuse from the elements.

The transmitter can modulate the carrier much faster than the rate which was used in the field tests. However, message transmission rates are currently limited by the real-time data processing capabilities of the receiver. Advances in receiver capabilities, such as implementation of nonlinear signal processing and/or machine learning methods on FPGA hardware, can relax this constraint. Using such an improved receiver would immediately enable increased message transmission rate and reduced energy consumption per symbol.

11.2 Future Work

11.2.1 Mechanical ELF Transmitters

It is interesting to note that the fundamental model constructed in Sect. 6.2 has no constraint which directly describes f_c. However, indirect influences have been explored. We might eliminate drag by placing the rotating system inside a vacuum chamber. How then, would power loss change? We might be able to pull a partial vacuum to reduce the influence of drag while maintaining the use of air bushings, though this would depend on the power required to maintain a steady vacuum. Channel conditions may influence the bounds for η_c in combination with detector sensitivity as shown in Chap. 10. Yet in a truer sense the ultimate optimization is in field transmission, for which we have simply expected reduced frequency to be the best solution. Future work should explore the cost-benefit analysis of increased transmit frequency with respect to receiver sensitivity and field generation efficiency, which may allow system performance to be improved by scaling the magnitude and frequency of the generated field. For example, recent work has demonstrated a 500,000 rpm (8.333 kHz) motor capable of generating 2 kW, which would allow electromechanical transmitters to be implemented well into the VLF range [5].

In addition, it is of interest to consider transmitter architectures that combine the electric motor and the source magnet(s). After all, an electric motor also uses magnets (either permanent magnets or electromagnets) to create the **B**-field required for generating torque. These magnets could potentially also serve as ELF field sources, thus considerably simplifying the transmitter design. However, in practice the **B**-field requirements for motors and field sources are significantly different: motors typically use well-confined fields from high-order multipoles to smoothly generate torque, unlike the unconfined dipole fields used by field sources. Thus, using such fields for motors is likely to result in decreased efficiency. Nevertheless, the resulting structures are likely to be much simpler to design from the mechanical point of view (e.g., by eliminating the need for bearings and shaft alignment), and are thus of interest for future research.

11.2.2 ELF Receivers

This book has primarily focused on the transmitter component of the ELF communication system. As a result, a mature ELF receiver has not been developed. Thus, a variety of improvements to the design shown in Chap. 9 are possible. For example, coil-based **B**-field sensors can be replaced with miniaturized alternatives, such as atomic magnetometers, to reduce the overall size and weight of the receiver. In addition, the performance of the receiver for a given sensor can be significantly improved by designing and implementing automated real-time data demodulation algorithms. Given the time-varying and non-Gaussian nature of noise in the ELF channel, improved channel coding, nonlinear signal processing, and machine learning methods can also be explored to minimize the bit error rate (BER) for a given received SNR [6, 7]. Fortunately, given the relatively low data rates involved, even sophisticated ELF signal processing algorithms can be readily implemented in real-time using an on-board FPGA.

References

1. Singh, A., & Sharma, S. P. (2016). Interpretation of very low frequency electromagnetic measurements in terms of normalized current density over variable topography. *Journal of Applied Geophysics, 133*, 82–91.
2. Inan, U. S. (1990). VLF heating of the lower ionosphere. *Geophysical Research Letters, 17*(6), 729–732.
3. Hayakawa, M. (2007). VLF/LF radio sounding of ionospheric perturbations associated with earthquakes. *Sensors, 7*(7), 1141–1158.
4. Madanayake, A., Choi, S., Tarek, M., Dharmasena, S., Mandal, S., Glickstein, J., & Sehirlioglu, A. (2017). Energy-efficient ULF/VLF transmitters based on mechanically-rotating dipoles. In *2017 Moratuwa Engineering Research Conference (MERCon)* (pp. 230-235). IEEE.
5. Islam, K., Choi, S., Hong, Y. K., & Kwak, S. (2021). Design of High-Power Ultra-High-Speed Rotor for Portable Mechanical Antenna Drives. *IEEE Transactions on Industrial Electronics.*
6. Bernstein, S., McNeill, D., & Richer, I. (1974). A signaling scheme and experimental receiver for extremely low frequency (ELF) communication. *IEEE Transactions on Communications, 22*(4), 508–528.
7. Ingram, R. (1984). Performance of the locally optimum threshold receiver and several suboptimal nonlinear receivers for ELF noise. *IEEE Journal of Oceanic Engineering, 9*(3), 202–208.

Spectral Content of Modulated PWM Signals

A

Equation 8.9 establishes the basic pulse train signal used for the modulation of motor speed. It is well known that the average "on" time of the pulse train controls the speed of the motor, and variance in this "on" time is used for modulation. In Chap. 8, the common approximation is used that low pass filtering in the electrical and mechanical response characteristics of the plant filters out all but the narrowband DC component of the signal. But what might be said of the actual frequency content of the PWM signal produced?

There are some notable exceptions to the common approximation, chiefly in high-power devices, engine-driven electronics, and audio equipment. High-power (>1 kW) devices use relatively low switching frequencies on the order of 1–2 kHz to minimize switching loss, as the absolute power loss is at least an order of magnitude greater than in a 100 W motorized device. Aviation electronics and power supplies operate on generators coupled to aircraft engines, and the switching frequency thus varies with engine speed (in the range of several 100 s–1000 Hz) [1].

Audio amplifiers are also in a position of intrigue. Not only is the switching frequency in both power supplies and digital amplifiers close to the signal range of interest, but also the final power amplifier stages driving speakers also qualify as high-power devices. Even some of the available consumer audio equipment delivers over 1 kW, for example, in a home theater.

Recent work has investigated how to model the frequency spectra of such signals and describe how higher order effects propagate through controlled systems [2]. Returning to the subject of AMEBA2, the transmitter operates for finite periods of time, transmitting sequences of packets having a deterministic length and limited dynamic range. Thus, the PWM signal generated by the transmitter is also of short duration and therefore finite in power. It might be very interesting to extract each component of the pulse train and analyze them individually [3].

J. S. Glickstein and S. Mandal, *Electromechanical Transmitters for ELF/VLF Radio*, Synthesis Lectures on Engineering, Science, and Technology, https://doi.org/10.1007/978-3-031-15123-1

A.1 Time-Domain Definition

Consider the rectangular pulse train of Sect. 8.2 having duty cycle $D = \frac{\tau}{T}$, where T is the period and τ is the "on" time over a given period. Equation 8.9 describes the pulse train as

$$p(t) = A \sum_{k=0}^{\infty} u(t - kT) - u(t - (k + D)T). \tag{A.1}$$

Of course D will in reality vary as a function of k and the summation over all k is in fact bounded by the length L of the transmission message. Thus, D is most appropriately a vector $D[k]$ containing L components, though for simplicity it will continue to be denoted by D. The gain A is of no interest to us and we will assume it to be unity for all further analysis.

As will soon be clear, it is convenient to treat the signal as the superposition of both periodic and aperiodic structures. That is,

$$p(t) = p_c(t) + p_m(t)$$

where the component $p_c(t)$ is a periodic continuous function and $p_m(t)$ is the modulating component of the signal. Suppose the periodic component is given by a square wave with initial condition $p_c(t) = 1$, taking the value

$$p_c(t) := \begin{cases} 1, & \text{if } t < \frac{1}{2}T \\ 0, & \text{if } t > \frac{1}{2}T \end{cases}$$

over each full period T. The square wave may also be expressed as a sum of unit step functions

$$p_c(t) = \sum_{k=0}^{L} u(t - kT) - 2u\left(t - \left(k + \frac{1}{2}T\right)\right) + u(t + (k + 1)T) \tag{A.2}$$

satisfying both initial condition and interval values. This expression is most useful as a means to describe the function of the modulating component,

$$p_m(t) = \sum_{k=0}^{L} 2u(t - (k + \frac{1}{2})T) - u(t - (k + D)T). \tag{A.3}$$

A.2 Frequency Domain

The carrier has an average value of zero and well-known Fourier series

$$p_c(t) = \frac{4}{\pi} \sum_{n=-\infty}^{\infty} \frac{1}{2n+1} \sin(2\pi(2n+1) f_0 t) \tag{A.4}$$

where f_0 is the carrier frequency.[1] The index of summation n corresponds to the $(2n+1)$-th harmonic and it is practical to note that for even large bandwidths the number of discrete values is relatively small, for example, if $f_0 = 20$ kHz, there are just 100 values in a 2 MHz two sided bandwidth.

It is possible to express Eq. A.4 as a sum of harmonics in terms of purely frequency by using the common Fourier transform

$$\mathscr{F}\{\sin(2\pi f_0 t)\} \mapsto \jmath \frac{\delta(f+f_0) - \delta(f-f_0)}{2}$$

where $\delta(x)$ is the *Kronecker delta* operating as a function [4]:

$$\delta(x) := \begin{cases} 1, & \text{if } x = 0, \\ 0, & \text{otherwise.} \end{cases}$$

The power spectrum $P_c(f)$ is then found to be

$$P_c(f) = \frac{2}{\pi} \sum_{n=0}^{N-1} \left[\frac{1}{2n+1} [\delta(f + (2n+1) f_0) - \delta(f - (2n+1) f_0)] \right]. \tag{A.5}$$

The modulator function is a different story. Concerning Eq. A.3, $p_m(t)$ is a series of time-shifted, translated unit step functions. The discrete Fourier transform of the unit step function is

$$\mathscr{F}\{u(t)\} \mapsto \frac{1}{\jmath 2\pi f} + \pi \delta(2\pi f) \tag{A.6}$$

which is applied using the time shift property $\mathscr{F}\{x(t-a)\} = \mathscr{F}\{x(t)\} e^{-\jmath 2\pi f a}$ to find

$$P_m(t) = \left(\pi \delta(2\pi f) + \frac{1}{\jmath 2\pi f}\right) \sum_{k=0}^{L} \left[2e^{-\jmath 2\pi f (k+\frac{1}{2})T} - e^{-\jmath 2\pi f (k+D)T} \right]. \tag{A.7}$$

Numerical evaluation of Eqs. A.5 and A.7 gives the spectra of Fig. A.1 for the carrier and modulator components. We find the carrier is expressed as only odd harmonics of f_0 and

[1] In Chap. 8, this frequency is denoted f_{PWM}.

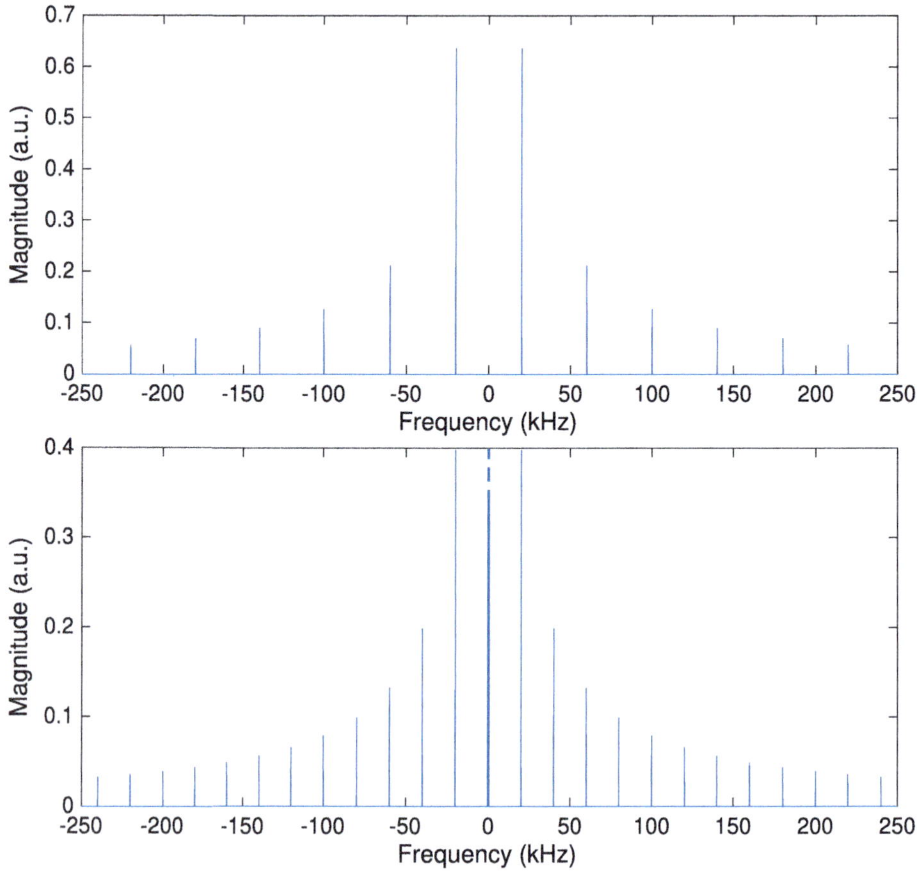

Fig. A.1 Distribution of spectral content in the components P_c (top) and P_m (bottom) for $f_0 =$ 20 kHz, $L = 5000$, and constant $D = 0.5$. The DC component of P_m is suppressed for visibility

the modulator contains both odd and even harmonics. The DC component $P_m(f = 0) \gg P_m(f \neq 0)$ is suppressed by indication with a dashed line that it continues to extend, and this is before consideration given to any filtering regimes. Note also that only magnitudes are shown in the figure.

The components P_c and P_m are not normalized to the same reference. However given P_c is independent of D, we may answer our original inquiry by observing P_m alone.

For example, a ramp where duty cycle is varied $D : 0.1 \rightarrow 0.9$ (from 10% to 90%) over 0.25 s contributes the spectral content of Fig. A.2 to the carrier. In the reverse case for $D : 0.9 \rightarrow 0.1$, the magnitude spectrum is the same however the phase of odd and even harmonics are changed, hence both real and imaginary components are shown in the figure.

Spectral content of a reversed modulation for $D : 0.9 \rightarrow 0.1$ is distinct from the previous case, shown in Fig. A.3. Thus, for any set of symbols $m_i \in M_t$, the spectral content of

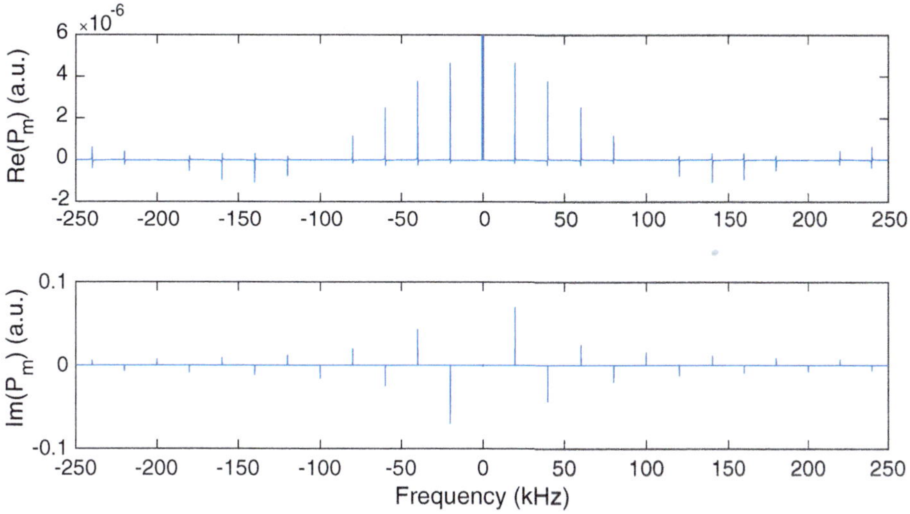

Fig. A.2 Spectral distribution of components in P_m for a scaled ramp function $D : 0.1 \rightarrow 0.9$ over $L = 5000$ cycles (0.25 s). The DC component is again suppressed for visibility

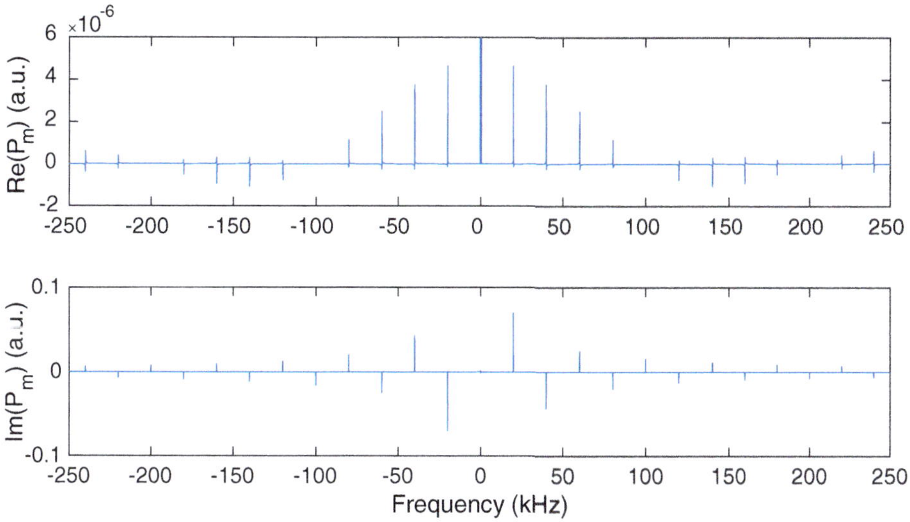

Fig. A.3 Reverse case of Fig. A.2, a scaled ramp function $D : 0.9 \rightarrow 0.1$ over $L = 5000$ cycles (0.25 s)

higher harmonics over the complete message is unique for each combination of symbols which it may contain. Furthermore, the modulation scheme employed will surely influence this fingerprint.

A.3 Direct Approach to Finding $P(t)$

If we are interested only in the spectral fingerprint of a message, it is not necessary to separate constant and modulating components into Eqs. A.5 and A.7. Instead one may consider a more direct route to $P(t)$ through applying the Fourier transform given in Eq. A.6 to our initial definition for the pulse train in Eq. A.1. Limiting the summation boundary to L, we find

$$P(t) = \left(\pi\delta(2\pi f) + \frac{1}{j2\pi f}\right)\sum_{k=0}^{L}\left[e^{-j2\pi fkT} - e^{-j2\pi f(k+D)T}\right] \tag{A.8}$$

which may be used to represent the complete spectral fingerprint for any arbitrary combination of symbols as a vector where $M_t \mapsto D$. For example, consider a message modulated by CP-FSK using a set having $M = 5$ levels. The controller is again set to $f_0 = 20$ kHz for the motor drive. If M_t is comprised of 10 randomly selected symbols from the set, the spectral distribution of one possible combination is shown in Fig. A.4. The result is produced under the trivial case assuming the speed response is linear and neglecting the control loop.

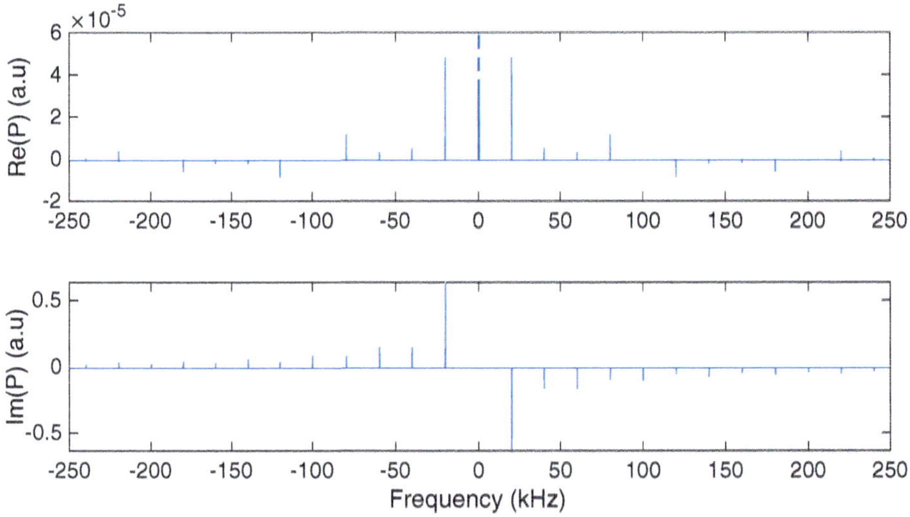

Fig. A.4 Unique spectral fingerprint of transmitter PWM waveform for a randomly selected series of 10 symbols from a set having $M = 5$ levels modulated by CP-FSK

In an experimental observation, the real response as observed near the transmitter (by magnetic, acoustic, or electrical means) will be influenced by the coupling between these domains and the digital controller. Predicted observations can be found by filtering $P(t)$ from Eq. A.8 based on the transfer function between these domains. Conclusively, higher order harmonics produced through modulation in the transmitter comprise a unique fingerprint with a distinct structure in the frequency domain. The impact of this observation is of potential interest regarding the security and anonymity of the transmitted message with respect to its source. The authors suggest that this topic be explored in future work.

References

1. Odavic, M., Sumner, M., Zanchetta, P., & Clare, J. C. (2009). A theoretical analysis of the harmonic content of PWM waveforms for multiple-frequency modulators. *IEEE Transactions on Power Electronics, 25*(1), 131–141.
2. John, M., & Mertens, A. (2020). Frequency-domain modeling of harmonic interactions in voltage-source inverters with closed-loop control. *Energies, 13*(21), 5823.
3. Song, Z., & Sarwate, D. V. (2003). The frequency spectrum of pulse width modulated signals. *Signal Processing, 83*(10), 2227–2258.
4. Woyczynski, W. A. (2011). *A first course in statistics for signal analysis*. Boston: Birkhäuser.

Glossary

ELF Extremely low frequency electromagnetic waves. As defined by the International Telecommunications Union (ITU), the ELF frequency range extends from 3–30 Hz. In this book, we use a broader definition of the ELF range (3 Hz–3 kHz) to simplify our notation and maintain consistency with common usage of the term in the atmospheric science community.

VLF Very low frequency electromagnetic waves. As defined by the International Telecommunications Union (ITU), the VLF frequency range extends from 3–30 kHz.

AMEBA Acronym for "A MEchanically Based Antenna", a term introduced by the U.S. Defense Advanced Research Projects Agency (DARPA) in 2017 to describe electromechanical ELF or VLF transmitters.

NdFeB Shorthand for Neodymium-Iron-Boron, the most widely used category of rare-earth magnet materials. NdFeB magnets are the strongest type of permanent magnets that are commercially available.

J. S. Glickstein and S. Mandal, *Electromechanical Transmitters for ELF/VLF Radio*, Synthesis Lectures on Engineering, Science, and Technology,
https://doi.org/10.1007/978-3-031-15123-1

Index

A
Adaptive tuning, 171
Air bearings, 81
Air filter, 110
Air friction, 84
All-mechanical transmitter, 4
AMEBA2, 69
Atomic magnetometers, 165
AWESOME, 57, 167

B
Ball bearings, 81
Bellows coupling, 83
BER analysis, 44
BLDC, 53
Brushless DC motor, 52
Burrows-Wheeler transform (BWT), 46

C
Chu-Harrington limit, 3
Closed-loop feedback system, 104
Command-line interface (CLI), 129
Conductive media, 1
Continuous-frequency FSK (CF-FSK), 41
Continuous phase modulation (CPM), 39
CP-FSK, 40
Critical speed, 28, 115
Cutler Naval Station, 8
Cuyahoga Valley National Park, 36

D
Data encoding, 62
DEFLATE, 39
Differential coding, 42
Direct memory access (DMA), 129
Distributed transmitters, 27

E
Electrical time constant, 75
ELF channels, 31
Encoder, 106
Extremely low frequency (ELF), 1

F
Fading, 45
Field programmable gate array (FPGA), 129
Field survey, 35
First-order system plus dead time (FOPDT), 156
Fixturing, 124
Flow multiplexer, 109
Forward counter, 137
Fundamental model, 71

J. S. Glickstein and S. Mandal, *Electromechanical Transmitters for ELF/VLF Radio*, Synthesis Lectures on Engineering, Science, and Technology,
https://doi.org/10.1007/978-3-031-15123-1

G
Gate driver, 120
Geomagnetic noise, 178
Grade N42, 51
Grade N52, 51

H
Half-bridge drivers, 106
Hard processor system (HPS), 129

J
Jaw coupling, 127

L
Lempel-Ziv algorithm, 39
Lock-in amplifier, 56
Low-noise amplifier (LNA), 55
Low-noise front-end, 171

M
Magnetic field sensors, 166
Mechanical time constant, 75
Memory-mapped parallel IO (PIO), 129
Message encoder, 152
Minimum energy configuration, 29
Multiple frequency-shift keying (MFSK), 40

N
Noiseless bandwidth adjustment, 171
Noncoherent detection, 71
Non-line-of-sight (NLOS), 54
Non-random data, 67
Notch filters, 57

O
Ohio Caverns, 189
Orthogonal frequency multiplexing (OFDM), 94

P
Phase-locked loop (PLL), 93
PID control, 145
Pneumatic system, 108
Portable transmitter, 181
Power-line harmonics, 178
Project ELF, 7
Project Sanguine, 7
Pulse width modulation (PWM), 135

Q
Q-damping, 171
Quasistatic approximation, 24

R
Reciprocal counter, 137
Reduced character set (RCS), 46
RF-denied environments, 3
Roller bearings, 81
Rotating electric dipole, 24
Rotating magnetic dipole, 23
Run-length encoding (RLE), 46

S
Schumann resonances, 33
Sea water, 3
Sendmail, 149
Sferics, 32
Shaft coupling, 82
Short-time Fourier transform (STFT), 59
Spider, 83, 127
Superconducting quantum interference device (SQUID), 165

U
Underground communication system, 16
Underwater communication systems, 13
Unit cells, 93
Universal asynchronous receiver/transmitter (UART), 131

V
Van Allen radiation belts, 32

W
Whistlers, 33
Wireless communication with submarines, 7

Z
ZEVS station transmitter, 12
Ziegler-Nichols, 156

The manufacturer's authorised representative in the EU is Springer Nature Customer Service Centre GmbH, Europaplatz 3, 69115 Heidelberg, Germany. If you have any concerns regarding our products, please contact ProductSafety@springernature.com

Printed and bound by CPI Group (UK) Ltd, Croydon, CR0 4YY
30/07/2026
02182060-0003